Lecture Notes in Social Networks

More information about this series at http://www.springer.com/series/8768

Mehmet Kaya • Jalal Kawash • Suheil Khoury
Min-Yuh Day

Editors

Social Network Based Big Data Analysis and Applications

 Springer

Editors
Mehmet Kaya
Department of Computer Engineering
Firat University
Elazig, Turkey

Jalal Kawash
Department of Computer Science
University of Calgary
Calgary, AB, Canada

Suheil Khoury
American University of Sharjah
Sharjah, United Arab Emirates

Min-Yuh Day
Tamkang University
Taipei, Taiwan

ISSN 2190-5428   ISSN 2190-5436   (electronic)
Lecture Notes in Social Networks
ISBN 978-3-030-08639-8   ISBN 978-3-319-78196-9   (eBook)
https://doi.org/10.1007/978-3-319-78196-9

Printed on acid-free paper

This Springer imprint is published by the registered company Springer International Publishing AG part
of Springer Nature.
The registered company address is: Gewerbestrasse 11, 6330 Cham, Switzerland

Preface

Social networks (SN) have brought an unprecedented revolution in how people interact and socialize. SN are used not only as a lifestyle but also in various other domains, including medicine, business, education, politics, and activism. The number of SN amounts to billions of users. At the beginning of 2016, Twitter claimed to have 313 million monthly active users. As of the third quarter of 2017, Facebook had slightly more than 2 billion monthly active users. Online social media (OSM), media produced by SN users, has offered a real and viable alternative to conventional mainstream media. OSM is likely to provide "raw," unedited information, and the details can be overwhelming with the potential of misinformation and disinformation. Yet, OSM is leading to the democratization of knowledge and information. OSM is allowing almost any citizen to become a journalist reporting on specific events of interest. This is resulting in unimaginable amounts of information being shared among huge numbers of OSM participants. For example, Facebook users are generating several billion "likes" and more than 100 million posted pictures in a single day. Twitter users are producing more than 6000 tweets per second. The size of the data generated presents increasing challenges to mine, analyze, utilize, and exploit such content. This book includes eleven contributions that examine several topics related to data analysis and social networks. Applications include sentiment dictionaries, malicious content identification, video recapping, cancer biomarkers, face detection, pattern detection, and cell phone subscription predictions. What follows is a quick summary of each of these chapters.

Nuno Guimarães, Luís Torgo, and Álvaro Figueira complement traditional sentiment dictionaries with a system for lexicon expansion, extracting and classifying domain- and time-specific terms with sentiment based on public opinion. Domain- and time-specific lexicons improve the performance of sentiment analysis methods on short informal texts, such as tweets. The proposed system can generate dictionaries, on a daily basis, to complement the more traditional sentiment lexicons.

Prateek Dewan, Shrey Bagroy, and Ponnurangam Kumaraguru address the issue of identifying malicious content on Facebook, such as publishing untrustworthy information, misleading content, adult and child unsafe content, and scams. The identified 627 malicious pages revealed through spatial and temporal analysis dominant presence of politically polarized entities engaging in spreading content from untrustworthy domains. Multiple supervised learning algorithms and multiple feature sets are evaluated, and they find that artificial neural networks trained on a fixed sized bag-of-words perform the best in identifying such malicious pages.

Automatic generation of video recaps and summaries is the subject of the chapter by Xavier Bost, Vincent Labatut, Serigne Gueye, and Georges Linarès. They propose narrative smoothing, a method for the extraction of dynamic social networks of video characters. They introduce an algorithm to estimate verbal interactions from a sequence of spoken segments. The data used are a corpus of 109 TV series episodes from three popular TV shows: Breaking Bad, Game of Thrones, and House of Cards.

Gabriela Jurca, Omar Addam, Jon Rokne, and Reda Alhajj study the assessment of candidates for academic positions or for promotion. They employ social network analysis and community detection to measure the influence and diversity of members, within the Department of Computer Science at the University of Calgary. Different measures between various ranks in the department are presented and discussed.

In another chapter, Gabriela Jurca, Omar Addam, Jon Rokne, and Reda Alhajj study biomarkers used to diagnose prostate cancer. They used text mining to provide a tool to examine whether biomarkers are emerging or decreasing in terms of publication popularity. They also provide a tool to examine the increasing or decreasing popularity of gene families with respect to prostate cancer research. Selected biomarkers which have been labeled as emerging in qualitative reviews are then evaluated.

The spread of influence in complex networks is the subject of the chapter by Arun Sathanur, Mahantesh Halappanavar, Yalin Sagduyu, and Yi Shi. They consider the problem of modeling the spread of influence and the identification of influential entities in a complex network with nodal activation, intrinsic or external through neighbors. They approach mining for the influential nodes through influence maximization. One of the findings is how influential content creators can drive engagement on social media platforms.

Yingbo Zhu, Zhenhua Huang, Zhenyu Wang, Linfeng Luo, and Shuang Wu revisit Spiral of Silence in the context of social networks with real information diffusion data. They analyze four information diffusion tree metrics: width, depth, message sentiment, and modularity. Based on Spiral of Silence, polarity prediction of users' review without considering semantic meaning of content is proposed and discovered. Their results indicate that opinions of people in propagation are impacted by the social environment. The Anti-Spiral of Silence is also found to play a significant role in leading rational public opinion and revealing truth in social networks.

Prediction of mobile service subscription types is entertained by Yongjun Liao, Wei Du, Márton Karsai, Carlos Sarraute, Martin Minnoni, and Eric Fleury, specifically the behavioral differences between prepaid and postpaid customers. The findings are used to provide methods that detect the subscription type of customers by using information about their personal call statistics and their egocentric networks. This allows this classification problem to be treated as a problem of graph labeling, which can be solved by max-flow, min-cut algorithms. The chapter also aims at inferring the subscription type of customers, using node attributes, and a two-ways indirect inference method based on observed hemophiliac structural correlations.

Konstantinos F. Xylogiannopoulos, Panagiotis Karampelas, and Reda Alhajj take on real-time detection of all repeated patterns in a big data stream. A new data structure is introduced: LERP Reduced Suffix Array with a new detection algorithm. This allows the detection of all repeated patterns in a string in a very short time. Specifically, their results show analysis of one million data points and a sliding window of groups of three subsequences of the same size simultaneously with detection in about 300 ms.

Cold start in a dating recommendation service is addressed by Mo Yu, Xiaolong Zhang, Dongwon Lee, and Derek Kreager. They approach this challenge by proposing a novel community-based recommendation framework. Detecting communities to which existing users belong and by matching new users to these communities, the proposed method improves on existing recommendation methods.

The last chapter by Salim Afra and Reda Alhajj studies the performance of face clustering approaches using different feature extraction techniques. Best practices for face recognition of terrorists and criminals are entertained. Performance evaluation for various feature extraction techniques and clustering algorithms using four datasets is also studied.

To conclude this preface, we would like to thank the authors who submitted papers and the reviewers who provided detailed constructive reports which improved the quality of the papers. Various people from Springer deserve great credit for their help and support in all the issues related to publishing this book.

Elazig, Turkey   Mehmet Kaya
Calgary, AB, Canada   Jalal Kawash
Sharjah, United Arab Emirates   Suheil Khoury
Taipei, Taiwan   Min-Yuh Day
November 2017

Contents

Twitter as a Source for Time- and Domain-Dependent Sentiment Lexicons

Nuno Guimarães, Luís Torgo, and Álvaro Figueira

Abstract Sentiment lexicons are an essential component on most state-of-the-art sentiment analysis methods. However, the terms included are usually restricted to verbs and adjectives because they (1) usually have similar meanings among different domains and (2) are the main indicators of subjectivity in the text. This can lead to a problem in the classification of short informal texts since sometimes the absence of these types of parts of speech does not mean an absence of sentiment. Therefore, our hypothesis states that knowledge of terms regarding certain events and respective sentiment (public opinion) can improve the task of sentiment analysis. Consequently, to complement traditional sentiment dictionaries, we present a system for lexicon expansion that extracts the most relevant terms from news and assesses their positive or negative score through Twitter. Preliminary results on a labelled dataset show that our complementary lexicons increase the performance of three state-of-the-art sentiment systems, therefore proving the effectiveness of our approach.

Keywords Lexicon expansion · Sentiment analysis · Social network applications

1 Introduction

A sentiment analysis task aims to, given a fragment of text, classify it with a score associated with a positive, neutral, or negative value. Early research has focussed on user reviews on online sites. However, the massive growth of social networks has provided a different source for sentiment analysis. This "boom" on the area was

N. Guimarães (✉) · Á. Figueira
CRACS - INESC TEC & University of Porto, Porto, Portugal
e-mail: nuno.r.guimaraes@inesctec.pt; arf@dcc.fc.up.pt

L. Torgo
LIAAD - INESC TEC & University of Porto, Porto, Portugal
e-mail: ltorgo@dcc.fc.up.pt

© Springer International Publishing AG, part of Springer Nature 2018   1
M. Kaya et al. (eds.), *Social Network Based Big Data Analysis and Applications*,
Lecture Notes in Social Networks, https://doi.org/10.1007/978-3-319-78196-9_1

caused mostly because of the way users share their opinion through short comments or texts, in several different domains. In addition, the large quantity of data available and the quickness of its extraction have recently promoted sentiment analysis to a "hot topic" research subject.

One of the key factors for a precise and correct sentiment analysis classification are sentiment lexicons or sentiment dictionaries. These consist in list of words, mainly adjectives and verbs, with an associated sentiment value (e.g., "beautiful: +2" and "bad: −1"). A basic example of a sentiment analysis procedure looks for all the words in the text that are within the dictionary. The sum of the values of those entries corresponds to the final sentiment score of the analyzed text.

Currently, there are several automatic and manually labelled sentiment dictionaries. However, the vast majority focus on opinion words such as adjectives like "beautiful" and "awful" or verbs like "lost" and "wins." Connotative words that are neither a verb nor an adjective, such as "cancer" and "terrorist," are not normally considered. Furthermore, when evaluating short informal texts, the absence of opinion words does not always imply the absence of sentiment. People often tend to use common knowledge to express an opinion without the use of sentiment words. For example, in the sentence "After Paris, Brussels. When it will end?" there is a clear presence of a negative sentiment (due to the terrorist attacks that happened in both cities [30]), but no opinion words to support it. This is because the opinion is expressed using facts regarding Paris which are normally common knowledge due to the impact that they had on news and the way the public reacted to it. In addition, time gains a specific importance when we are dealing with this type of sentiment analysis with absence of opinion words. In fact, for most people, the fragment above would have no meaning by itself or sentiment associated prior to the terrorist attacks [11].

Besides time, these words must also have a domain associated to them. For example, if we consider the text fragment "listening to Prince, I still can believe it." The meaning of "Prince" in this particular sentence is specific to the entertainment/music domain (since it is the name of a well-known musician) and not the more general concept (i.e., a member of royalty).

Therefore, it is plausible to say that the sentiment of terms, like the ones mentioned above, may vary through time and according to the domain. This is more visible if we consider entities like persons or organizations. Again, in the example of the word "Paris," it is fair to presume that the sentiment of the word was different before and shortly after the terrorist attacks on the city.

Therefore, our research hypothesis states: "Can domain and time specific lexicons improve the performance of sentiment analysis methods on short informal texts?". With that goal in mind, we propose a system that automatically extracts and classifies domain- and time-specific terms with sentiment based on public opinion. This system can "return" dictionaries, on a daily basis, to complement more traditional sentiment lexicons.

2   Related Work

There have been several approaches to the creation and/or expansion of sentiment dictionaries. We can classify them as manually labelled, thesaurus-based, and corpus-based approaches.

Manually labelled sentiment dictionaries rely on human annotators to assess the score on each entry. The author in [27] selected a set of words from several affective word lists (like ANEW [4]), added slang and obscene terms, and manually labelled with a sentiment score ranging from −5 to 5. Another work [19] takes a similar approach. The authors create a manually labeled sentiment dictionary by inspecting already well-established lexicons and adding acronyms and slang words. Then, recurring to Amazon Mechanical Turk [1], they assess each word sentiment using ten independent workers and a careful quality control on the data extracted. Finally they combine this lexicon with a rule-based system that takes into consideration negations ("not good"), degree modifiers ("very good"), punctuation, and capitalization to outperform seven state-of-the-art sentiment lexicons. Another approach [25] classifies words with emotion (e.g., anger, sadness, and happiness) and polarity (positive/negative). The terms were extracted from a combination of The Macquarie Thesaurus [5], General Inquirer [34], and WordNet Affect Lexicon [43].

Corpus-based approaches rely on the use of a text corpus already labeled (e.g., in a semi-supervised or unsupervised fashion) to create or expand a sentiment lexicon. One of the first works conducted in this area was using a small seed lexicon and conjunctions (such as "and" or "but") to determine the polarity of adjectives [16]. A more recent work [15] presents a methodology to create Twitter corpus-based lexicon. The process consists in the extraction of tweets with only a happy (:) or :-)) or sad (:(, :-( ) emoticon. Then, the assumption is that tweets with a smiling emoticon correspond to positive tweets and with a sad emoticon to negative ones. Finally, the corpus is divided (considering the emoticons) and the most frequent words in each are included in the lexicon with a positive or negative value. A similar approach is presented in [24]. However, instead of emoticons, the tweet retrieval process is done with emotion hashtags such as "#angry" and "#happy." The lexicon evaluates each word with six emotions and positive and negative sentiment. The authors in [31] use a different approach. Using a small seed lexicon, they are capable to extract sentiment words and features from reviews and expand domain-specific lexicons. The method consists in defining direct and indirect relations between words in sentences (with the help of a POS-tagger) and then, using a set of rules, extract sentiment and feature words. To assign the polarity of the sentiment words extracted, the authors rely on observations such as assign the same polarity for a feature in a review and the same polarity for a sentiment word in a domain-specific corpus.

Finally, thesaurus-based approaches use word resources like WordNet [10] for expanding a small sentiment lexicon (seed lexicon). WordNet is a lexical database that includes nouns, verbs, and adjectives grouped by synonyms sets. In adjectives, there is also a connection between antonyms. This is particularly useful for expanding sentiment lexicons. As an example, SentiWordNet uses this feature to expand a seed lexicon by assigning the same polarity to synonyms and the opposite to the antonyms [8]. Several other studies [18, 20] use WordNet to expand sentiment or create sentiment lexicons, making it one of the most used resources for the creation or expansion of dictionaries.

In other work [26], the authors expand a sentiment lexicon in a two-step procedure. First, they generate the seed lexicon by using eleven affix patterns (e.g., the affix "dis" is used to detect the pair dishonest-honest) and then they use The Macquarie thesaurus to expand the previous defined dictionary. This method is different from the most since it generates and classifies automatically the seed lexicon. The results show that the number of correct entries is far superior to the ones provided by SentiWordNet.

Nevertheless, none of these approaches considers the use of relevant (domain- and time-dependent) terms, which may be crucial for the correct polarity assessment on short informal texts—ultimately, constitutes the motivation for this work.

3   System Workflow

As it was mentioned before, it is the goal of this work to assess if time- and domain-specific sentiment lexicons can improve the state-of-the-art sentiment analysis methods. In this section, we describe the workflow of the system developed to extract these lexicons automatically.

Our system is divided into two main components: the term extraction and term sentiment evaluation. First we select six of the most common news categories to extract our time- and domain-dependent terms: "world," "entertainment," "politics," "sports," "health," "technology," and "business." Next, using different news sources, we crawl the headlines on a daily basis to retrieve the more relevant terms on each domain. Then, we use those terms as queries to extract a corpus of tweets for each one of them. Finally, using sentiment analysis procedures on tweets referring to that term, we assess the public opinion of it. Our hypothesis is that the overall sentiment of the *corpus* extracted using the term as a keyword corresponds to the sentiment of the term.

3.1   Terms Extraction

To extract the more relevant terms in each domain, we create a corpus of headlines from several different news sources. The number of sources in each domain ranges between 9 and 14. We limited our news sources research to the English language and whose origin countries are the United States or included in the United Kingdom. In fact, a survey puts the United States and UK as the two most influential countries in the world according to several different factors [17]. Therefore, we argue that international media coverage is bigger in these countries and consequently, public opinion data should also be vast and easier to acquire using terms from these geographical sources. The sources used were CNN, BBC, The Economist, The Wall Street Journal, ABC News, CBS News, The Washington Post, NBC, The Guardian, Reuters, Yahoo News, Sky News, Daily Mail, The New York Times, Financial Times, Forbes, and MedicineNet.

For each domain corpus, we remove punctuation and impose lowercase. Then, we build three lists by extracting all unigrams, bigrams, and trigrams in order of frequency. Through experimentation, we realize that, most of the times, terms above trigrams were unique (in other words, they only occur in one headline) so we discard them.

Next, we perform a series of text filtering. We exclude both verbs and adjectives from the lists using OpenNLP Part of Speech Tagger [2]. In addition, we also exclude terms that are within the domain and are not subject to public opinion (e.g., "soccer" in sports or "film" in entertainment) recurring to the word lists provided by Oxford's Topic Dictionaries [28]. Furthermore, we also exclude possible sentiment words using AFINN lexicon [27]. This way, some subjective adjectives or verbs that could pass the OpenNLP classifier are left out. Finally we removed words that were duplicated in plural form ("syrian"/"syrians"), and lemmatized when in the presence of an apostrophe ("Clinton"/"Clinton's").

The POS-Tagger filter is only applied to unigrams whereas the other filters are used in all lists, since terms with two or more words already imply a certain context. The last filter is applied to the three lists at the time when the sample of tweets for each term is extracted. Through experimentation, we defined a threshold of 33% on the minimum sample of tweets to be retrieved. Consequently, terms below that minimum are excluded. Since we are searching for an exact match on the queries, incomplete or irrelevant terms are unlikely to reach the minimum number of tweets. The workflow of this component is presented in Fig. 1.

3.1.1 Term Extraction Evaluation

To evaluate the term extraction component, we conduct an experimental survey to determine if the terms were up to date and belonged to the domain from where they were retrieved. The survey was conducted during the time period of 2 days (16 and 17 of March 2016). The question asked was "Considering the present time (and current news), does the term x fits the domain y?" where x and y were replaced randomly by the entries extracted from our system. The possible responses were "Yes," "No," and "I don't know" in case the user was unfamiliar with the term.

The survey was shared among social networks and university students. We do not restrict the number of terms that each user could evaluate being only limited to the full extension of the term list extracted. In addition, the terms are extracted from a

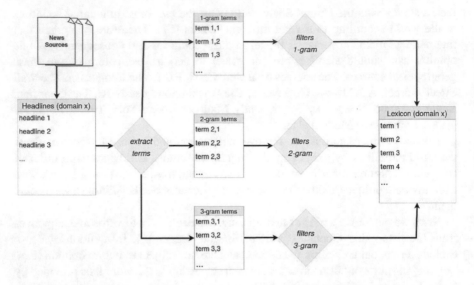

Fig. 1 Term extraction component

global term list and assigned to each individual user in a consecutive way. Therefore, with this approach we will have approximately the same number of evaluations by term.

A total of 1414 entries were classified by approximately 60 different users consisting mostly of university students. We discarded all results whose response was "I don't know" which correspond to 5.5% of all evaluations. Furthermore, we only considered terms that had at least three evaluations and we consider our groundtruth the majority of the evaluations.

Our results show an accuracy of 90.9% on the fitness of the domain and time. In 4.2% of the terms, consensus among evaluators was not achieved and in 4.9% our term extraction feature failed to correctly assess the domain or time of the term. Although more or less expected (since we are retrieving terms from categories such as news headlines), these results provide strong empirical evidence for our term selection method.

3.2 Term Sentiment Evaluation

The second component of our system determines the sentiment or public opinion of the extracted terms. The majority of corpus-based approaches rely only on Twitter to extract the terms and classify them with sentiment. To the best of our knowledge, this is the first system for lexicon expansion that extracts terms from one source (news headlines) and assesses the sentiment on other (Twitter).

To determine the sentiment of each term extracted, we use the term as a key word in the Twitter API [42]. We then retrieve 100 tweets related to it. This number was achieved by experimental procedures which took into account the restrictions imposed by the Twitter API as well as the time to classify each tweet with sentiment.

We also impose some restrictions on the tweets extracted for each term. Since we want to keep the sentiment updated, we only retrieve tweets posted in the same day as the term extraction procedure and in the English language. In addition, we use the parameters provided by the Twitter REST API [41] to retrieve the most recent tweets. Furthermore, in order to avoid extracting posts by news sources (since we want to analyze the sentiment exposed by common users and not by news media) we do not extract tweets that contain an external link. This is due to the fact that the majority of Twitter accounts that belong to the news industry refer to their web page in each news post (so the user can read the full article).

As soon as the tweet term corpus is built, we applied some cleaning procedures to it and begin the sentiment analysis in each tweet. For that purpose, we built an ensemble system (ENS17) which takes into account a selection of sentiment analysis methods to improve the inter-domain performance. The methods/sentiment dictionaries used were the following:

- **AFINN**: Twitter-based sentiment lexicon expanded from ANEW [4]. It contains words that are frequently used in this social network such as Internet slang and offensive words [27].
- **Emolex**: Manually created emotion lexicon using crowdsourcing. The terms were extracted from a combination of The Macquarie Thesaurus [5], General Inquirer [34], and WordNet Affect Lexicon [43]. Although the words were classified with emotion and polarity, only the second was used for this method [25].
- **EmoticonDS**: It is a lexicon created using a corpus-based approach. The method consists in the extraction of tweets with only a happy ("*:)*" or "*:-)*") or sad ("*:(*", "*:-(*") emoticon. Then, the assumption is that tweets with a smiling emoticon correspond to positive tweets and with a sad emoticon to negative ones. Finally, the corpus is divided considering the emoticons and the most frequent words in each division are included in the lexicon [15].
- **Happiness Index**: Uses words from ANEW that were manually classified with a 1–9 happiness scale. To assess the sentiment, this method considers that positivity is achieved when the happiness value for a tweet is between 6 and 9 whereas negativity is between 1 and 4. Tweets with no words associated or with happiness value 5 are considered neutral [13].
- **MPQA (or Opinion Finder)**: It is a machine-learning model to detect subjectivity and consequently, the polarity of a sentence based on sentiment clues [45]. Since each sentence can have more than one sentiment clue, this method considers the sum of them as the final sentiment score.
- **NRC Hashtag**: Uses the same concept as EmoticonsDS, although, instead of emoticons, the tweet retrieval process is done with emotion hashtags such as "#angry" and "#happy." The lexicon evaluates each word with six different emotions and positive and negative sentiment [24].

- **Opinion Lexicon**: Extracts and classifies opinion words from a corpus of reviews to build a lexicon. Uses a thesaurus-based approach and a seed lexicon of 30 words as a starting point [18].
- **SANN**: Uses the sentiment lexicon of MPQA along with polarity shifters, negation, and amplifiers to build a sentence-level sentiment classifier. It was originally used in user comments present in Ted Talks videos [29].
- **Sasa**: It is a supervised method based on a Naive-Bayes approach. It uses the unigram features of each tweet. This method was originally used to detect sentiment on tweets in real time during the U.S. 2012 election [44].
- **SenticNet**: Assigns sentiment to common sense concepts to achieve a semantic sentiment analysis approach rather than the most common sentence level [6].
- **Sentiment140 Lexicon**: Is a corpus-based sentiment lexicon extracted from the tweets provided in [12]. It has similarities with the NRC Hashtag method for lexicon extraction and practically equal to the EmoticonDS (only in a different corpus).
- **SentiStrength**: Combines a manually annotated sentiment lexicon, machine-learning algorithms, and other important features like negation words and repeated punctuation for sentiment enhancement. It provides the best results in gold-standard tweet datasets [38–40].
- **SentiWordNet**: Is a lexical resource which provides all WordNet entries with a positive, negative, or neutral polarity. A short lexicon consisting of seven positive terms and seven negative terms were used. Next, a dictionary-based approach was used on WordNet, with a limited reach on each word (meaning that each seed lexicon entry should not expand by synonyms or antonyms more than k times). Finally, all the classified terms are used as training data on a supervised model to assign a score to the remaining ones [3].
- **SoCal**: Uses a sentiment dictionary and features like negation and amplification words. The authors claim that the dictionaries used are robust through several Mechanical Turk evaluations [35].
- **Stanford Adapter**: Uses a deep learning scheme more concretely a Recursive Neural Tensor Network to determine the sentiment at a sentence level. This method provides a differentiating feature which is the order of the words in the sentence is taken into account for sentiment assessing [33].
- **Umigon Adapter**: Is a system designed specifically for tweets sentiment analysis. It is a dictionary-based approach that has characteristics like the detection of smileys and onomatopes (e.g., "yeeeeeaaaaaah"), hashtag evaluation (e.g., detecting negative sentiment in #notverygood), and decomposition of the tweet in n-grams (to be able to distinguish "good" from "not good") [21]
- **Vader**: Directed for microblogging sentiment analysis, Vader uses sentiment lexicons of words, smileys, and Internet acronyms and slang, validated by human annotators. Furthermore, it also evaluates the impact of punctuation and uppercase words using Mechanical Turk. All this is combined in a rule-based system with polarity shifters and trigram analysis (for negation detection and amplification words) [19].

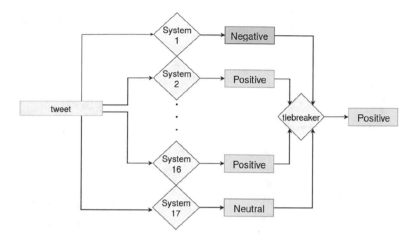

Fig. 2 Ensemble system example

To facilitate the building of the ensemble we used the iFeel framework, which allows the selection of specific sentiment analysis methods [23, 32]. The decision-making procedure on the score returned is done by majority voting. When a tie occurs, the rules are the following:

– When there is a tie between positive and negative classes, the neutral sentiment is returned.
– When there is a tie between the neutral class and other class, the other class is returned.
– Since there are 17 sentiment systems, a 3-way tie is not possible. However, assuming different setups in terms of ensemble composition are possible, we define that in this case, the neutral value is returned.

An example of the behaviour of the ensemble system can be seen in Fig. 2.

It is important to point out that some of the methods are solely based on the use of dictionaries and thus it is necessary to specify how the classification of the text will be done. Taking this into account, for the lexicon only approaches (AFINN, Emolex, EmoticonDS, NRC Hashtag, Opinion Lexicon, Sentiment 140, and SentiWordNet), iFeel uses Vader rule-based system to push forward the performance of these lexicons [32].

In previous work [14], we conclude that term classification using three classes (Negative/Neutral/Positive) was needed. Therefore, to determine the sentiment of the term based on the *corpus* of tweets extracted we used the following formula:

$$\text{score}_c = \frac{\text{number of tweets classified as } c}{\text{total number of tweets}}$$

where c is the respective sentiment class (Negative, Neutral, or Positive).

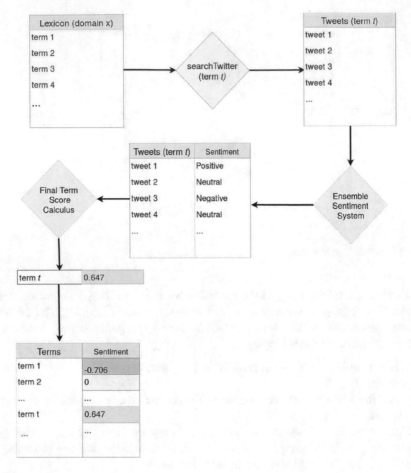

Fig. 3 Term sentiment evaluation

This formula gives us the confidence of the sentiment in each one of the classes. Consequently, the sentiment of the term is assessed using the class with the maximum score value. If the neutral class has the maximum score, we ignore the confidence value and assign a 0 score to that term. Otherwise, we use the confidence value of the class returned by the formula (multiplying it by −1 in the cases where the maximum score belongs to the negative class). This way, the scores for the created lexicons will range from [−1, 1].

Figure 3 represents the workflow of the term sentiment evaluation component.

3.2.1 Ensemble System Evaluation

Since an accurate tweet sentiment analysis is essential for the results of our system, in this section we compare our approach against the individual state-of-the-art methods which comprise our ensemble system, on a sample of different domain tweet datasets. The datasets used were extracted from CrowdFlower Data for Everyone Library [7] and included a set of tweets referring to the 2016 GOP debate (GOP), Google self-driving cars (SDC), Coachella line-up announcement (COACH), United States airlines (USAIR), and the Deflategate scandal (NFL). Since we want a good performance across all domains and classes, we select an equal number of entries in each dataset and balance the three classes available. Therefore, each of the previous mentioned datasets has 1200 tweets (400 positive, 400 neutral, and 400 negative). We assess our ensemble results in terms of accuracy and average $F1$-score in different classes and different domains. First, we compare our ensemble system with the top three more accurate systems in each domain and using the aggregation of all datasets, represented by the "Average" column. The results are presented in Table 1. When compared with each stand-alone system, the ENS17 ensemble is in the top three most accurate in almost all datasets (it fails in the COACH dataset, although the difference is 0.1%).

This ensemble does not achieve the best score in any of the datasets with the exception of the NFL and GOP. We assume that this is due to the large numbers of jokes in the tweets included [37] and political irony, which may make difficult the classification task in the individual systems. However, if we look at the accuracy across all domains (in other words, the average accuracy in all datasets), the ENS17 outperforms the best individual systems. We stress out that the "Average" column represents the values for the top three individual systems that perform better in the concatenation of all datasets, and not the average of the remaining columns.

A similar analysis can be done separating the accuracy in Negative, Neutral, and Positive classification. With this purpose, we will use the top three systems that are more accurate in all domains (i.e., the individual systems that were selected for the "Average" column in Table 1). These systems are AFINN, SentiStrength, and Umigon. The table regarding class accuracy can be examined in Table 2.

Table 1 Comparison of ensemble method against the more accurate individual systems in each domain

	Dataset accuracy						
	GOP	SDC	APPLE	USAIR	COACH	NFL	Average
Top individual systems (on each dataset)							
First system	49.0	**53.0**	**69.5**	**62.0**	**46.5**	35.2	48.8
Second system	48.1	51.3	61.3	59.3	46.1	34.1	48.3
Third system	47.5	47.5	54.6	58.4	45.3	33.4	48.0
Ensemble system							
ENS17	**51.0**	51.6	60.4	60.0	45.2	**45.2**	**52.2**

The results are presented in Table 1 and the best score for each system in highlighted in bold

Table 2 Comparison of ensemble method against the most accurate individual systems

	Class accuracy (%)			
	Negative	Neutral	Positive	Total
Best overall individual systems (using accuracy as metric)				
Umigon	33.8	**77.3**	35.1	48.8
SentiStrength	**36.8**	63.2	44.7	48.3
AFINN	36.3	59.3	48.3	48.0
Ensemble system				
ENS17	35.8	72.3	**48.6**	**52.2**

The results are presented in Table 2 and the best score for each system in highlighted in bold

Table 3 Comparison of ensemble method against the top individual systems (according to $F1$-metric) in each domain

	Dataset $F1$-score (%)						
	GOP	SDC	APPLE	USAIR	COACH	NFL	Average
Top individual systems (on each dataset)							
First system	48.6	**52.3**	**69.1**	**61.9**	**44.6**	32.7	47.8
Second system	47.5	50.6	60.9	58.6	43.5	31.8	47.2
Third system	45.8	46.6	55.0	57.6	42.3	30.1	47.6
Ensemble system							
ENS17	**50.3**	50.8	60.4	59.4	42.2	**42.2**	**51.5**

The results are presented in Table 3 and the best score for each system in highlighted in bold

As we can observe, regarding classes, ENS17 is always in the top three systems when comparing with the most accurate individual systems. Furthermore, it achieves the highest accuracy value on the positive class. Since the datasets are balanced in the number of entries and elements in each class, it's no wonder that the all-classes accuracy values are the same as the total accuracy values in all datasets. Since accuracy values can sometimes be misleading [36], we perform the same analysis using the average $F1$-score in each domain. Therefore, selecting the top three systems according to the average $F1$-score and assessing the same metric with our method results in the values presented in Table 3.

Once again it is clear that our ensemble system performs well enough to be in the top three systems using $F1$-score in each dataset. Therefore, it is no surprise that, when considering all datasets, ENS17 achieves an average $F1$-score superior to each of the individual systems.

Finally, we take a closer look on the performance of the class classification using the concatenation of all datasets and the $F1$-score metric. Results are provided in Table 4. It is easily noticeable that the ensemble system outperforms the individual systems, therefore proving the validity of our approach.

Table 4 Comparison of ensemble method against the top individual systems (according to $F1$-metric) in each class

	Class $F1$-score (%)			
	Negative	Neutral	Positive	Average
Best overall individual systems (using F1 measure as metric)				
SentiStrength	44.0	51.2	48.2	47.8
AFINN	44.3	50.2	48.3	47.6
Umigon	41.8	54.7	45.2	47.2
Ensemble system				
ENS17	**46.6**	**55.6**	**52.2**	**51.5**

The results are presented in Table 4 and the best score for each system in highlighted in bold

4   Lexicons Evaluation

The final stage of this work is to determine if the dictionaries built with the described system can improve the sentiment analysis task for short 'informal' texts. To answer our research question, we used a dataset that contains posts and comments from Facebook and tweets from September 7 to September 14 2016, evaluated with sentiment on CrowdFlower. The dictionaries from our system were retrieved on September 5th. This way, we guarantee that the tweets and Facebook posts used to create the dictionaries were not included in the dataset where we performed the evaluation but are close enough so the public opinion on the terms extracted does not fade or change substantially.

4.1   Dataset Description

As it was already mentioned, the dataset combines three types of short informal texts: Facebook posts, Facebook comments, and tweets. The Facebook posts and comments were retrieved from the top most popular pages in different categories from the United States according to the LikeAlyzer tool [22]. For each post on the defined time interval, we extracted up to a maximum of 20 comments (order by the Facebook "ranked" metric [9]). From that extraction, a sample of 1000 comments and 3995 posts were sent to CrowdFlower for evaluation.

Regarding the tweets extraction, relevant topics (which appeared on recent news) were used on the Search API. To retrieve a large number of tweets in different domains, we used the terms we knew it would generate opinion tweets. Therefore, some keywords used as queries were evaluated with sentiment in our lexicons. Consequently to avoid biased results, we excluded those terms from the dictionaries. The key words used as queries were the following: "terrorism," "refugees," "elections," "paralympic," "champions league," "emmys," and "wall street."

For each keyword, 714 tweets were extracted forming a total of 4998 tweets. Concatenating this data with the one extracted from Facebook, we have a final dataset of 9993 entries of short informal texts for evaluation.

The survey on CrowdFlower consisted in two sentiment questions. The first was "The sentiment expressed in this text is:" To answer, the workers had a Likert scale ranging from 1 to 5 and labeled from "very negative" to "very positive." The second was a follow-up question that stated: "Choose (from the provided text) the word that best supports your previous answer". Our goal was to lead the worker to take a more careful decision and to justify it. Finally, since we want to assess the impact of our complementary lexicon on improving the accuracy on subjectivity texts, we exclude the entries classified as neutral (since they are very likely to be factual) of our dataset. This left us with a dataset containing 5090 entries.

4.2   Evaluation on Nonfactual Texts

We select AFINN [27], UMIGON [21], and SentiStrength [40] as the sentiment methods to complement with our lexicons since (1) they are well-known systems in the state of the art of sentiment analysis and (2) in the tests previously mentioned they were the systems that individually performed better in terms of accuracy and average $F1$-score.

To decide on which lexicon to use in each entry of the dataset, we need to fit each text in one of the domains previously defined (world, sports, entertainment, politics, business, technology, and health). In other words, we need to assign a domain for each entry of the dataset. In this experiment, we used the frequency of words on the text that appear on Oxford's Topic Dictionaries [28] combined with the dictionaries generated by our system to assess its domain. For the entries where no domain was found, we assigned the "world" value.

Finally, we scale the sentiment classification on CrowdFlower to Negative or Positive values to match our methods scales. The results of our experiment are presented in Table 5.

The addition of the lexicons outputted by our system improved the tested methods in both accuracy and average $F1$-score. Umigon is the system that benefits the most on the addition of these lexicons and AFINN the less. The average accuracy improvement is around 1.87%, whereas F-measure is 1.51%.

We can conclude that, although is not a major difference between both sentiment dictionary approaches (traditional and traditional + expanded), it is a steady improvement since it is consistent across all three analyzed systems.

Table 5 Variation between the sentiment systems with and without the expanded lexicons

Sentiment system	Accuracy %	Average $F1$%
AFINN	+1.12	+0.48
SentiStrength	+1.36	+1.43
Umigon	+3.14	+2.63

Table 6 Variation between the sentiment systems with and without the expanded lexicons with sentiment justification word in the expanded lexicons

Sentiment system	Accuracy %	Average $F1\%$
AFINN	+2.23	+2.31
SentiStrength	+23.13	+9.81
Umigon	+24.11	+12.55

The main reason why our approach does not improve on a greater scale the results from traditional sentiment analysis lexicons is due to the specificity of the problem we are trying to solve. In fact, if we go further in our analysis and restrict our dataset to the entries whose response to the question "Choose (from the provided text) the word that best supports your previous answer" was included in our expanded sentiment lexicon, we can really tackle the problem we are trying to solve. The filtered dataset contains 215 entries and results of these specific cases can be consulted in Table 6.

Although we are "forcing" that the word for the sentiment justification is present in our dictionary (and therefore imposing the condition that it will be used for the text sentiment evaluation), this analysis intends to show that, in specific cases of subjective short informal texts where the argument to assess the sentiment is not on traditional lexicons, using our system can result in a reasonable improvement. In fact, SentiStrength and Umigon have an accuracy boost superior to 20%, whereas their $F1$-score increases 9.81% and 12.55%, respectively. This demonstrates that it is important not only to consider our system sentiment dictionaries but also that our term sentiment analysis is capable of accurately classifying the terms. Therefore, the results show that the addition of our lexicons improves the performance of state-of-the-art sentiment systems. Furthermore, they make a major difference when we analyze subjective texts whose sentiment is not determined by opinion words (such as verbs and adjectives) included in traditional sentiment lexicons.

5 Conclusion and Future Work

In this work, we studied the influence of public opinion for the task of assessing a positive/negative sentiment in subjective short informal texts (like tweets, posts, or comments).

We built a framework capable of extracting and assessing the polarity score of the most relevant domain- and time-dependent terms. This system consisted in an extraction procedure (that relies on news headlines to retrieve relevant terms) and on an ensemble tweet sentiment classifier (combining 17 state-of-the-art sentiment analysis methods to analyze tweets regarding the terms). The final output is seven different sentiment dictionaries that are retrieved on a daily basis.

Next, we complement three state-of-the-art sentiment systems (AFINN, UMIGON, and SentiStrength) with the dictionaries outputted from our method. We tested our approach on a sample of tweets, Facebook posts, and comments with positive or negative polarity and whose value was manually assigned recurring to CrowdFlower platform.

The results achieved indicate a coherent improvement in all methods. In addition, when the term for assessing the sentiment is not included in sentiment dictionaries, the importance of our domain- and time-specific lexicons increases significantly, proving that our approach can increment the performance of sentiment methods in these specific cases. These results allow us to conclude that, although our lexicons try to solve a specific problem, they are effective on that task and do not compromise the performance of traditional sentiment analysis methods.

It is important to notice that the lexicons generated by this framework do not replace traditional sentiment lexicons. The goal is to complement them, by having domain and time sentiment terms attached to state-of-the-art methods. This approach goes against what is normally proposed in the area, since the majority of works have focus in building sentiment lexicons and compare them with state-of-the-art methods.

Therefore, we do believe that our framework can keep up and complement even the more recent proposals on sentiment methods. However, our lexicons integration is simpler in rule-based approaches, since that supervised methods would require repeating the learning phase with the extended lexicons.

For this reason, in future work, it would be interesting to discover the "expiration date" of the lexicons generated. In other words, to analyze how the time difference between the extraction of the lexicons and the source of the text affects the performance on sentiment analysis tasks. This can be particularly useful to integrate the dictionaries created by our framework in supervised methods without having to create new models on, for example, a daily basis.

In addition, although the results achieved are promising, in future work to conduct our evaluation we intend to use domain-specific tweet datasets (instead of using an automatic domain disambiguation). We do believe that with a domain already defined, the performance of the lexicons will increase. We also plan to further extend our system by adding a geographical component to the dictionaries generated. We intend to use news sources as well as tweets from specific countries for determining the effectiveness of our method in a more narrow scope evaluation. We also aim to extend our lexicons in more domains. This way, we can increase the number of terms and cover a broader area of short informal texts to be analyzed.

Acknowledgements This work is supported by the ERDF European Regional Development Fund through the COMPETE Programme (operational programme for competitiveness) and by National Funds through the FCT (Portuguese Foundation for Science and Technology) within project Reminds/UTAP-ICDT/EEI-CTP/0022/2014.

References

1. Amazon: Amazon mechanical turk. https://www.mturk.com/mturk/welcome (2016). Accessed 21 Aug 2016
2. Apache: Opennlp: http://opennlp.apache.org (2010). Accessed 07 Mar 2016
3. Baccianella, S., Esuli, A., Sebastiani, F.: Sentiwordnet 3.0: an enhanced lexical resource for sentiment analysis and opinion mining. In: Calzolari, N., Choukri, K., Maegaard, B., Mariani, J., Odijk, J., Piperidis, S., Rosner, M., Tapias, D. (eds.) Proceedings of the Seventh International Conference on Language Resources and Evaluation (LREC'10), Valletta. European Language Resources Association (ELRA), Paris (2010)
4. Bradley, M.M., Lang, P.J.: Affective norms for English words (ANEW): stimuli, instruction manual, and affective ratings. Technical Report, Center for Research in Psychophysiology, University of Florida, Gainesville (1999)
5. Butler, S.: The Macquarie thesaurus/[general editor] J.R.L. Bernard, new budget edn. Herron Publications, West End (1987)
6. Cambria, E., Olsher, D., Rajagopal, D.: Senticnet 3: a common and common-sense knowledge base for cognition-driven sentiment analysis. In: Proceedings of the Twenty-Eighth AAAI Conference on Artificial Intelligence, AAAI'14, pp. 1515–1521. AAAI Press, Palo Alto (2014)
7. Crowdflower: Data for everyone. http://www.crowdflower.com/data-for-everyone/ (2016). Accessed 10 Apr 2016
8. Esuli, A., Sebastiani, F.: Sentiwordnet: a publicly available lexical resource for opinion mining. In: Proceedings of the 5th Conference on Language Resources and Evaluation (LREC'06), pp. 417–422 (2006)
9. Facebook: Facebook Graph API. https://developers.facebook.com/docs/graph-api/reference/v2.8/object/comments (2016). Accessed 19 Oct 2016
10. Fellbaum, C. (ed.): WordNet: An Electronic Lexical Database. MIT Press, Cambridge (1998)
11. Foster, A.: Terror attacks timeline: from Paris and Brussels terror to most recent attacks in Europe. http://www.express.co.uk/news/world/693421/Terror-attacks-timeline-France-Brussels-Europe-ISIS-killings-Germany-dates-terrorism (2016). Accessed 21 Aug 2016
12. Go, A., Bhayani, R., Huang, L.: Twitter sentiment classification using distant supervision. CS224N Project Report, Stanford 1, 12 (2009)
13. Gonçalves, P., Araújo, M., Benevenuto, F., Cha, M.: Comparing and combining sentiment analysis methods. In: Proceedings of the First ACM Conference on Online Social Networks, pp. 27–38. ACM, New York (2013)
14. Guimaraes, N., Torgo, L., Figueira, A.: Lexicon expansion system for domain and time oriented sentiment analysis. In: Proceedings of the 8th International Joint Conference on Knowledge Discovery, Knowledge Engineering and Knowledge Management - Volume 1: KDIR (IC3K 2016), pp. 463–471 (2016)
15. Hannak, A., Anderson, E., Barrett, L.F., Lehmann, S., Mislove, A., Riedewald, M.: Tweetin' in the rain: exploring societal-scale effects of weather on mood. In: Proceedings of the 6th International Conference on Weblogs and Social Media (2012)
16. Hatzivassiloglou, V., McKeown, K.R.: Predicting the semantic orientation of adjectives. In: Proceedings of the 35th Annual Meeting of the Association for Computational Linguistics and Eighth Conference of the European Chapter of the Association for Computational Linguistics, ACL'98, pp. 174–181. Association for Computational Linguistics, Stroudsburg (1997)
17. Haynie, D.: The U.S. and U.K. are the world's most influential countries, survey finds. www.usnews.com/news/best-countries/best-international-influence (2015). Accessed 23 May 2016
18. Hu, M., Liu, B.: Mining and summarizing customer reviews. In: Proceedings of the Tenth ACM SIGKDD International Conference on Knowledge Discovery and Data Mining, KDD'04, pp. 168–177. ACM, New York (2004)

19. Hutto, C.J., Gilbert, E.: Vader: a parsimonious rule-based model for sentiment analysis of social media text. In: Adar, E., Resnick, P., Choudhury, M.D., Hogan, B., Oh, A.H. (eds.) ICWSM. The AAAI Press, Menlo Park (2014)
20. Kim, S.-M., Hovy, E.: Determining the sentiment of opinions. In: Proceedings of the 20th International Conference on Computational Linguistics, COLING'04. Association for Computational Linguistics, Stroudsburg (2004)
21. Levallois, C.: Umigon: sentiment analysis for tweets based on lexicons and heuristics. In: Proceedings of 7th International Workshop on Semantic Evaluation (SemEval 2013), Atlanta (2013). Zenodo
22. LikeAlyzer: Likealyzer: analyze and monitor your Facebook pages. http://likealyzer.com/ (2016). Accessed 21 Sept 2016
23. Messias, J., Diniz, J.P., Soares, E., Ferreira, M., Araújo, M., Bastos, L., Miranda, M., Benevenuto, F.: Towards sentiment analysis for mobile devices. In: Kumar, R., Caverlee, J., Tong, H. (eds.) 2016 IEEE/ACM International Conference on Advances in Social Networks Analysis and Mining, ASONAM 2016, San Francisco, 18–21 August 2016, pp. 1390–1391. IEEE Computer Society, New York (2016)
24. Mohammad, S.M.: #emotional tweets. In: Proceedings of the First Joint Conference on Lexical and Computational Semantics - Volume 1: Proceedings of the Main Conference and the Shared Task, and Volume 2: Proceedings of the Sixth International Workshop on Semantic Evaluation, SemEval'12, pp. 246–255. Association for Computational Linguistics, Stroudsburg (2012)
25. Mohammad, S.M., Turney, P.D.: Emotions evoked by common words and phrases: using Mechanical Turk to create an emotion lexicon. In: Proceedings of the NAACL HLT 2010 Workshop on Computational Approaches to Analysis and Generation of Emotion in Text, CAAGET'10, pp. 26–34. Association for Computational Linguistics, Stroudsburg (2010)
26. Mohammad, S., Dunne, C., Dorr, B.: Generating high-coverage semantic orientation lexicons from overtly marked words and a thesaurus. In: Proceedings of the 2009 Conference on Empirical Methods in Natural Language Processing: Volume 2 - Volume 2, EMNLP'09, pp. 599–608. Association for Computational Linguistics, Stroudsburg (2009)
27. Nielsen, F.Å.: A new ANEW: evaluation of a word list for sentiment analysis in Microblogs. In: Proceedings of the ESWC2011 Workshop on "Making Sense of Microposts": Big Things Come in Small Packages, pp. 93–98 (2011)
28. Oxford: Oxford Learner's Dictionaries topic dictionaries. http://www.oxfordlearnersdictionaries.com/topic/ (2016). Accessed 03 Jul 2016
29. Pappas, N., Popescu-Belis, A.: Sentiment analysis of user comments for one-class collaborative filtering over ted talks. In: 36th ACM SIGIR Conference on Research and Development in Information Retrieval. ACM, New York (2013)
30. Phipps, C.: Brussels: Islamic state launches attacks on airport and station – as it happened. https://www.theguardian.com/world/live/2016/mar/22/brussels-airport-explosions-live-updates (2015). Accessed 28 Sept 2016
31. Qiu, G., Liu, B., Bu, J., Chen, C.: Opinion word expansion and target extraction through double propagation. Comput. Linguist. 37(1), 9–27 (2011)
32. Ribeiro, F.N., Araújo, M., Gonçalves, P., Gonçalves, M.A., Benevenuto, F.: Sentibench-a benchmark comparison of state-of-the-practice sentiment analysis methods. EPJ Data Sci. 5(1), 1–29 (2016)
33. Socher, R., Perelygin, A., Wu, J.Y., Chuang, J., Manning, C.D., Ng, A.Y., Potts, C.: Recursive deep models for semantic compositionality over a sentiment Treebank. In: Proceedings of the Conference on Empirical Methods in Natural Language Processing (2013)
34. Stone, P.J., Dunphy, D.C., Smith, M.S., Ogilvie, D.M.: The General Inquirer: A Computer Approach to Content Analysis. MIT Press, Cambridge (1966)
35. Taboada, M., Brooke, J., Tofiloski, M., Voll, K., Stede, M.: Lexicon-based methods for sentiment analysis. Comput. Linguist. 37(2), 267–307 (2011)
36. Tenuto, J.: Classification accuracy is not enough: more performance measures you can use. http://machinelearningmastery.com/classification-accuracy-is-not-enough-more-performance-measures-you-can-use/ (2014). Accessed 20 Aug 2016

37. Tenuto, J.: #deflategate was just a chance for us to make some really bad jokes. https://www. crowdflower.com/deflategate-sentiment/ (2015). Accessed 12 Aug 2016
38. Thelwall, M.: Heart and soul: sentiment strength detection in the social web with Sentistrength. In: Proceedings of CyberEmotions, p. 1–14 (2013)
39. Thelwall, M., Buckley, K., Paltoglou, G., Cai, D., Kappas, A.: Sentiment in short strength detection informal text. J. Am. Soc. Inf. Sci. Technol. **61**(12), 2544–2558 (2010)
40. Thelwall, M., Buckley, K., Paltoglou, G.: Sentiment strength detection for the social web. J. Am. Soc. Inf. Sci. Technol. **63**(1), 163–173 (2012)
41. Twitter: Rest API documentation. https://dev.twitter.com/rest/public (2015). Accessed 19 Oct 2015
42. Twitter: Twitter Developers . https://dev.twitter.com/ (2016). Accessed 08 Mar 2016
43. Valitutti, R.: Wordnet-affect: an affective extension of wordnet. In: Proceedings of the 4th International Conference on Language Resources and Evaluation, pp. 1083–1086 (2004)
44. Wang, H., Can, D., Kazemzadeh, A., Bar, F., Narayanan, S.: A system for real-time twitter sentiment analysis of 2012 U.S. presidential election cycle. In: Proceedings of the ACL 2012 System Demonstrations, ACL'12, pp. 115–120. Association for Computational Linguistics, Stroudsburg (2012)
45. Wilson, T., Hoffmann, P., Somasundaran, S., Kessler, J., Wiebe, J., Choi, Y., Cardie, C., Riloff, E., Patwardhan, S.: Opinionfinder: a system for subjectivity analysis. In: Proceedings of HLT/EMNLP on Interactive Demonstrations, HLT-Demo'05, pp. 34–35. Association for Computational Linguistics, Stroudsburg (2005)

Hiding in Plain Sight: The Anatomy of Malicious *Pages* on Facebook

Prateek Dewan, Shrey Bagroy, and Ponnurangam Kumaraguru

Abstract Facebook is the world's largest Online Social Network, having more than one billion users. Like most social networks, Facebook is home to various categories of hostile entities who abuse the platform by posting malicious content. In this chapter, we identify and characterize Facebook *pages* that engage in spreading URLs pointing to malicious domains. We revisit the scope and definition of what is deemed as "malicious" in the modern day Internet, and identify 627 *pages* publishing untrustworthy information, misleading content, adult and child unsafe content, scams, etc. We perform in-depth characterization of *pages* through spatial and temporal analysis. Upon analyzing these *pages*, our findings reveal dominant presence of politically polarized entities engaging in spreading content from untrustworthy web domains. Studying the temporal posting activity of *pages* reveals that malicious *pages* are 1.4 times more active daily than benign *pages*. We further identify collusive behavior within a set of malicious *pages* spreading adult and pornographic content. Finally, we attempt to automate the process of detecting malicious Facebook *pages* by extensively experimenting with multiple supervised learning algorithms and multiple feature sets. Artificial neural networks trained on a fixed sized bag-of-words perform the best and achieve a maximum ROC area under curve value of 0.931.

Keywords Facebook · Online Social Networks · Malicious URLs

This chapter is an extended version of the paper titled "Hiding in Plain Sight: Characterizing and Detecting Malicious Facebook Pages" previously accepted at ASONAM 2016.

P. Dewan (✉) · S. Bagroy · P. Kumaraguru
IIIT-Delhi, Delhi, India
e-mail: prateekd@iiitd.edu.in; shrey14099@iiitd.ac.in; pk@iiitd.ac.in

© Springer International Publishing AG, part of Springer Nature 2018
M. Kaya et al. (eds.), *Social Network Based Big Data Analysis and Applications*,
Lecture Notes in Social Networks, https://doi.org/10.1007/978-3-319-78196-9_2

1 Introduction

Online Social Networks (OSNs) are an integral part of the modern Internet. Users around the world use OSNs as primary sources to consume news, updates, and information about events around the world. However, given the enormous volume and veracity of content on social networks, it is hard to moderate all content that is generated and shared on OSNs. This enables hostile entities to generate and promote all sorts of malicious content (including scams, fake information, adult content, etc.) and pollute the information stream for monetary gains, or to compromise system reputation. Such activity not only degrades user experience, but also violates the terms of service of OSN platforms.

Researchers have studied and proposed automated techniques to identify malicious user accounts on OSNs [2, 16, 17, 26, 38]. Most of these techniques have a restricted focus of malicious content, which is limited to promotional posts, duplicate bulk messages, campaigns, phishing, and malware. However, with the advent of OSNs and Web 2.0, the scope of what is deemed as "malicious" on the Internet has evolved. Facebook, for example, has established community standards to safeguard users against nudity, hate speech, etc. [13], and considers any pages, groups, or events that confuse, mislead, surprise, or defraud people as abusive [12]. In a recent study, we discovered the presence of a similar set of malicious Facebook *pages* accounting for over 30% of malicious posts in our dataset, and have not been studied in detail [8]. Security experts and news sources have also acknowledged the presence of malicious *pages* on Facebook, set up intentionally to spread fraudulent claims, scams, and other types of malicious content. A group of scammers, for example, set up a fake British Airways *page*, offering free flights to customers for a year (Fig. 1). The *page* asked users to *share* a photo, *like* the *page* and leave a *comment* to claim their free flights.[1] In another similar incident, an international football player's name was used as bait to set up a Facebook *page*, and users were asked to sign a fake petition.[2]

In addition to scams and fake information, researchers have also identified and studied the spread of rumors (which are a class of untrustworthy information) on Facebook [15]. In case of events like earthquakes, rumors on OSNs have been observed to contribute to chaos and insecurity in the local population [29]. Facebook also faced criticism because of presence of fake news and polarized politics on the platform during the recent presidential elections in the USA [18]. Such instances highlight a new and emerging class of malicious content, which is much harder to identify using automated means and hasn't been widely explored in literature. It is thus crucial to identify and control the spread of untrustworthy and fake information and minimize adverse real world impact.

---

[1] https://grahamcluley.com/2015/09/british-airways-isnt-giving-away-free-flights-year-facebook-scam/.

[2] http://www.marca.com/2014/07/18/en/football/barcelona/1405709402.html.

Fig. 1 Fake British Air Facebook page offering free flights for a year in return for liking, commenting on, and sharing their post

In this chapter, we identify and characterize a set of 627 Facebook *pages* that published one or more malicious URLs (URLs that point to webpages comprising malicious content) in their most recent 100 posts. We expand on our characterization study, feature sets, and supervised learning experiments from the previous version [9]. We focus our analysis on *pages* which spread untrustworthy information, hate speech, nudity, misleading claims, etc., that are deemed as malicious by the community standards [13] and "*Page* Spam" definitions [12] established by Facebook. We use our labeled dataset to train and evaluate multiple supervised learning models to automate the process of identifying malicious Facebook *pages*. We extract a total of 96 features from *page* information and posts published by the *pages*. Further, we train and evaluate supervised learning models using a bag-of-words obtained using the textual content published by these *pages*. Our broad findings are as follows:

- **Politically polarized malicious entities**: We identified and manually verified the presence of numerous politically polarized entities, which dominated our dataset of malicious *pages*, and published URLs from untrustworthy web domains.
- **Malicious *pages* were more active**: We found that malicious *pages* were more active (in terms of posting) than benign *pages*; the number of malicious *pages* that were active daily was over three times the number of benign *pages* that were active daily.
- **Malicious *pages* showed collusive behavior**: We found presence of collusive behavior within malicious *pages* in our dataset; malicious *pages* engaged in promoting (*liking, commenting on*, and *sharing*) each other's content.
- **Malicious and benign *pages* had similar temporal behavior**: We performed a longitudinal study over a period of 1 year, by capturing daily snapshots of malicious and benign *pages* in our dataset, and found minimal statistically significant difference between the two types of *pages* in terms of temporal behavior.
- **Artificial neural networks outperformed all other algorithms**: Artificial neural networks trained on a bag-of-words outperformed all other supervised learning algorithms for automatic detection of malicious *pages*, achieving an area under the ROC curve value of 0.9. Grid search experiments help improve the performance further, attaining a maximum ROC AUC value of 0.931.

The rest of the chapter is structured as follows. Related work is discussed in Sect. 2. Section 3 gives the background and scope of our research, and explains the data collection process. Characterization and analysis of malicious *pages* make up Sect. 4. We present the results of our supervised learning experiments in Sect. 5. Section 6 discusses the implications and limitations of our results. We conclude and discuss the future directions of our work in Sect. 7.

2   Related Work

Detecting malicious content beyond spam and phishing, like fake reviews, link farming, social spam, etc. has received some attention by the research community [23]. These pieces of research have highlighted the surge of non-traditional spam on OSNs in general, and have also proposed network based techniques to combat suspicious behavior [22]. These techniques are however restricted by the scarcity of data available in practice.

Social Spam Detection   Fake reviews and opinionated spam has been prevalent and well studied in the ecommerce domain. Mukherjee et al. employed human labelers to identify groups of fake reviewers on Amazon. Authors proposed several behavior features derived from collusion among fake reviewers, and proposed a relation-based ranking model for automatic detection of spammer groups. Due to the human labeling required for generating ground truth, this method was limited in terms of scalability and generalizability [30]. Owing to this drawback, authors proposed an unsupervised Bayesian inference model to identify spammers [31]. Although this technique was novel and shown to be effective, the features used by authors (for example, extreme rating, rating deviation) cannot be ported directly to the social network domain.

Similar attempts were made by Jindal and Liu [24] and Lim et al. [27], both of which exploited review ratings and feedbacks from reviews to identify fraudulent activity using ranking and classification techniques under supervised settings. Manual effort needed for these techniques, in terms of ground truth generation and evaluation respectively, was a limitation in both of these attempts.

Akoglu et al. [3] and Fei et al. [14] proposed network based approaches to identify fraudulent reviewers by modeling the reviewer network as a graph. Although these approaches performed reasonably well, the lack of network level data in practice on OSNs like Facebook makes it difficult for porting and evaluating such approaches for identifying malicious entities on Facebook. Also, network level approaches help identify fraudulent users on the network, while the focus of our study is more on the content.

Ratkiewicz et al. studied astroturf political campaigns on Twitter using supervised learning. Similar to previous approaches, this work also required human labeling for ground truth generation [35]. Authors followed up on this approach and presented Truthy, a web service that tracked political memes in Twitter and helped detect astroturfing, smear campaigns, and other misinformation in the context of US political elections [34]. This research is closely aligned to our work. We overcome the drawbacks and limitations introduced because of human labeling by leveraging crowdsourced services like Web of Trust to extract ground truth for our study.

Malicious Content on Facebook   Multiple researchers have studied and proposed techniques to detect malicious content on Facebook. Gao et al. presented an initial study to quantify and characterize spam campaigns launched using accounts on Facebook [16]. They studied a large anonymized dataset of 187 million asynchronous "wall" messages between Facebook users, and used a set of automated techniques to detect and characterize coordinated spam campaigns. Authors of this work relied on URL blacklists to detect spam URLs and concentrated on spam, phishing, and malware. Following up their work, Gao et al. presented an online spam filtering system that could be deployed as a component of the OSN platform to inspect messages generated by users in real time [17].

In an attempt to protect Facebook users from malicious posts, Rahman et al. designed a social malware detection method which took advantage of the social context of posts [33]. Authors were able to achieve a maximum true positive rate of 97%, using a SVM based classifier trained on seven features. These features included no. of likes, no. of comments, no. of wall posts, no. of news feed posts, message similarity score, spam keyword score, and a boolean feature capturing whether a URL present in a post was obfuscated. Similar to Gao et al.'s work [16], this work was also targeted at detecting spam campaigns.

Stringhini et al. [38] utilized a honeypot model to collect information about spammers on Facebook. They created and monitored a honey profile for over 1 year, and manually identified 173 spam profiles among a total of 3831 friendship requests they received. Studying the spam profiles, authors developed six features, viz. follower-friend ratio, message similarity, URL ratio, no. of friends, no. of messages sent, and friend choice. Authors trained a Random Forest classifier on 173 spam and 827 legitimate profiles, and reported a false positive rate of less than 3% for both Facebook and Twitter networks. Ahmed et al. presented a Markov Clustering (MCL) based approach for the detection of spam profiles on Facebook. Authors crawled the public content posted by 320 handpicked Facebook users, out of which, 165 were manually identified as spammers. Authors then extracted three features from these profiles, viz. Active friends, *Page* Likes, and URLs to generate a weighted graph, which served as input to the Markov Clustering model [2].

Most aforementioned research relied on URL blacklists to identify ground truth spam, phishing, and malware, and tried to identify patterns which could be used to design effective measures to curb the spread of spam on OSN platforms. However, fewer attempts have been made to go beyond the traditional spam, phishing, and malware, and address other classes of malicious content on OSNs which include untrustworthy content, hate and discrimination, etc. that are non-trivial to identify through automated means. There has been some research in the space of identifying credible content on Twitter [5, 21], but state-of-the-art techniques proposed by researchers to detect content credibility have not been able to achieve the degree of efficiency that has been achieved in detecting spam, phishing, and malware.

3 Scope and Data Collection

Facebook (unlike Twitter, Instagram, etc.) poses a restriction on the number of connections a user can have (max. 5000 friends), and provides *pages* to enable large following for celebrities, groups, businesses, etc. A Facebook *page* can have multiple administrators managing the *page* under the same name, without the audience knowing. This allows *pages* to have a higher degree of interaction with its audience and keeping it more active as compared to a single user profile. Facebook *pages* are an important and integral part of the Facebook ecosystem that offer a free platform for promotion of businesses, brands, and organizations.[3] From an attacker's perspective, Facebook *pages* are potentially lucrative tools to gather large audiences and target all of them at once. Our past research has shown greater

---

[3]https://www.facebook.com/help/174987089221178.

participation of Facebook *pages* in posting malicious URLs as compared to posting benign URLs [8]. Such inflated malicious activity and reach of Facebook *pages* make them a vital aspect to study in detail.

3.1 Scope

The definition and scope of what should be labeled as "malicious content" on the Internet has been constantly evolving since the birth of the Internet. Researchers have been studying malicious content in the form of spam and phishing for over two decades. With respect to Online Social Networks, state-of-the-art techniques have become efficient in automatically detecting spam campaigns [16, 45] and phishing [1] without human involvement. However, new classes of malicious content pertaining to appropriateness, authenticity, trustworthiness, and credibility of content have emerged in the recent past. Some researchers have studied these classes of malicious content on OSNs and shown their implications in the real world [5, 19, 20, 29]. All of these studies, however, resorted to human expertise to identify untrustworthy and inappropriate content and establish ground truth, due to the absence of efficient automated techniques to identify such content. We aim to study a similar class of malicious content pertaining to trustworthiness and appropriateness in this work, which currently requires human expertise to identify. In particular, we look at Facebook *pages* that generate content deemed as malicious by Facebook's community standards and definitions of *"Page* Spam." Facebook defines *"Page* Spam" as *pages* that *confuse, mislead, surprise, or defraud people* [12]. Additionally, *pages* that are misleading, deceptive, or otherwise misrepresent the prize or any other aspect of promotion are considered as *"Page* Spam." Facebook has also established community standards to protect users against nudity, hate speech, violence and graphic content, fraud, sexual violence, etc. [13].

3.2 Establishing Ground Truth

Given the complex definition of malicious content for the scope of our study, there exist no accurate detectors for establishing ground truth. Detectors such as URL blacklists (Google Safebrowsing, PhishTank, SURBL, SpamHaus, etc.) used for identifying malicious content are restricted to identifying classical threats like phishing, malware, etc. In order to obtain ground truth for the malicious content we aim to study, we resorted to a crowdsourced approach. Crowdsourcing techniques have been shown to perform well for establishing ground truth for complex and subjective aspects of OSN content such as credibility [5, 19]. For our study, we used the Web of Trust (WOT) API [43]. WOT leverages crowdsourcing to collect ratings and reviews from millions of users who rate and comment about websites, based on their personal experiences. This crowdsourced, community based mechanism enables WOT to protect users against threats that only the human eye can spot such as scams, unreliable web stores, misleading websites, nudity, and questionable content, which largely overlaps with Facebook definitions of spam. To the best of our knowledge, WOT is one of the only services which covers the broader definition of malicious content that is required for our study.

We understand that WOT ratings are obtained through crowdsourcing, and may seem to suffer from biases. However, WOT states that in order to keep ratings more reliable, the system tracks each user's rating behavior before deciding how much it trusts the user. In addition, the meritocratic nature of WOT makes it far more difficult for spammers to abuse. This approach is similar to other crowdsourcing services like Amazon's Mechanical Turk[4] and CrowdFlower,[5] which have been widely used in OSN research (as discussed above). To further increase the confidence in the ratings, we used conservative thresholds for confidence values associated with the reputation scores. We discuss these thresholds in more detail in Sect. 3.3.

3.3 Dataset

We collected an initial dataset of 4.4 million public posts published by 390,246 unique *pages* and 2,983,707 unique users on Facebook between April 2013 and July 2014, using Facebook's post search API. These posts were collected by using event related search keywords belonging to 17 real world events that took place in the aforementioned time frame. All events we picked for our analysis made headlines in international news. To maintain diversity, we selected events covering various domains of news events like political, sport, natural hazards, terror strikes, and entertainment news. For all 17 events, we started data collection from the time the event took place, and stopped about 2 weeks after the event ended. Table 1 shows the detailed description of this data.

We queried the WOT API for domain reputations of all URLs present in the 4.4 million posts (480,407 unique URLs), and identified 10,341 posts containing one or more malicious URLs (4622 unique URLs). We will discuss the exact definition of "malicious" later in the section. These 10,341 posts containing malicious URLs originated from 1557 *pages* and 5868 users.

To capture a more recent view of the 1557 *pages* posting malicious URLs, we re-queried the Graph API and collected their *page* information in August 2015. We also collected 100 most recent posts (or all posts, whichever was smaller) published by these *pages* using the */page-id/posts* endpoint of the Graph API[6] along with all *likes*, *comments*, and *shares* on these posts. Collecting posts along with their likes, comments, and shares is a computationally expensive and time-consuming task. Due

---

[4]http://mturk.com/.

[5]http://crowdflower.com/.

[6]https://developers.facebook.com/docs/graph-api/reference/page/feed.

Table 1 Event name, keywords used as search queries, number of posts, and description for the 17 events in our dataset

Event (*keywords*)	# posts	Description
Missing Air Algerie Flight AH5017 (*ah5017; air algerie*)	6767	Air Algerie flight 5017 disappeared from radar 50 min after take off on July 24, 2014. Found crashed near Mali; no survivors
Boston Marathon Blasts (*prayforboston; marathon blasts; boston marathon*)	1,480,467	Two pressure cooker bombs exploded during the Boston Marathon at 2:49 pm EDT, April 15, 2013, killing 3 and injuring 264
Cyclone Phailin (*phailin; cyclonephailin*)	60,016	Phailin was the second-strongest tropical cyclone ever to make landfall in India on October 11, 2013
FIFA World Cup 2014 (*worldcup; fifaworldcup*)	67,406	20th edition of FIFA world cup, began on June 12, 2014. Germany beat Argentina in the final to win the tournament
Unrest in Gaza (*gaza*)	31,302	Israel launched Operation Protective Edge in the Hamas-ruled Gaza Strip on July 8, 2014
Heartbleed bug in OpenSSL (*heartbleed*)	8362	Security bug in OpenSSL disclosed on April 1, 2014. About 17% of the world's web servers found to be at risk
IPL 2013 (*ipl; ipl6; ipl2013*)	708,483	Edition 6 of IPL cricket tournament hosted in India, April–May 2013
IPL 2014 (*ipl; ipl7*)	59,126	Edition 7 of IPL cricket tournament jointly hosted by United Arab Emirates and India, April–May 2013
Lee Rigby's murder in Woolwich (*woolwich; londonattack*)	86,083	British soldier Lee Rigby attacked and murdered by Michael Adebolajo and Michael Adebowale in Woolwich, London on May 22, 2013
Malaysian Airlines Flight MH17 shot down (*mh17*)	27,624	Malaysia Airlines Flight 17 crashed on 17 July 2014, presumed to have been shot down, killing all 298 on board
Metro-North Train Derailment (*bronx derailment; metro north derailment; metronorth*)	1165	A Metro-North Railroad Hudson Line passenger train derailed near the Spuyten Duyvil station in the New York City borough of the Bronx on December 1, 2013. Four killed, 59 injured
Washington Navy Yard Shootings (*washington navy yard; navy yard shooting; NavyYardShooting*)	4562	Lone gunman Aaron Alexis killed 12 and injured 3 in a mass shooting at the Naval Sea Systems Command (NAVSEA) headquarters inside the Washington Navy Yard in Washington, DC on Sept. 16, 2013
Death of Nelson Mandela (*nelson; mandela; nelsonmandela; madiba*)	1,319,745	Nelson Mandela, the first elected President of South Africa, died on December 5, 2013. He was 95
Birth of the fist Royal Baby (*RoyalBabyWatch; kate middleton; royalbaby*)	90,096	Prince George of Cambridge, first son of Prince William, and Catherine (Kate Middleton), was born on July 22, 2013
Typhoon Haiyan (*haiyan; yolanda; typhoon philippines*)	486,325	Typhoon Haiyan (Yolanda), one of the strongest tropical cyclones ever recorded, devastated parts of Southeast Asia on Nov. 8, 2013
T20 Cricket World Cup (*wt20; wt2014*)	25,209	Fifth ICC World Twenty20 cricket competition, hosted in Bangladesh during March–April, 2014. Sri Lanka won the tournament
Wimbledon Tennis 2014 (*wimbledon*)	2633	128th Wimbledon Tennis championship held between June 23, and July 6, 2014. Novak Djokovic from Serbia won the championship

to limited time and resources, we had to restrict our dataset to 100 most recent posts. We then looked up the WOT API for all URL domains present in the most recent 100 posts, and found that 627 *pages* published one or more malicious URLs. This exercise of rescanning the 1557 *pages* eliminated those *pages* which had not shown malicious activity in the recent past (recent 100 posts), and could be deemed as non-malicious for our study. For the rest of the chapter, we use the remaining 627 *pages* as our dataset of malicious *pages*.

According to its documentation,[7] the WOT API returns a reputation score for a given domain. Reputations are measured for domains in several *components*, for example, trustworthiness. For each {domain, component} pair, the system computes two values: a *reputation* estimate and the *confidence* in the reputation. Together, these indicate the amount of trust in the domain in the given component. A *reputation* estimate of below 60 indicates *unsatisfactory*. The WOT browser add-on requires a confidence value of ≥ 10 before it presents a warning about a website. We tested the domain of each URL in our dataset for *Trustworthiness* and *Child Safety* components. For our analysis, a URL was marked as malicious if both the aforementioned conditions were satisfied ($reputation < 60; confidence \geq 10$). In addition to reputations, the WOT rating system also computes categories for websites based on votes from users and third parties. We marked a URL as malicious if it fell under the *Negative* or *Questionable* category (Table 2). We used the same approach previously to develop techniques for automatic identification of individual malicious posts on Facebook [8].

We also drew an equal random sample of 1557 *pages* from the benign *pages* in our dataset of 4.4 million posts, which had not posted any malicious URLs during our initial data collection phase (April 2013–July 2014). Similar to our approach for identifying malicious *pages*, we re-queried the Graph API and collected the *page* information along with the most recent 100 posts (including their *likes*, *comments*, and *shares*) published by these *pages*. We found 1278 *pages* which published no malicious URLs in their most recent 100 posts. These 1278 *pages* made up our dataset of benign *pages*. The remaining *pages* were either deleted, migrated, or had been merged with other *pages*. Table 3 shows the descriptive statistics of all Facebook *pages* in our dataset.

Table 2 Category labels and descriptions returned by WOT API

Category	Description
Negative	Malware, viruses, poor customer experience, phishing, scam, potentially illegal, adult content
Questionable	Misleading claims, unethical, privacy risks, suspicious, hate, discrimination, spam, potentially unwanted programs, ads, pop-ups, incidental nudity, gruesome/shocking

Source: WOT API Wiki (https://www.mywot.com/wiki/API)

---

[7]https://www.mywot.com/wiki/API.

Table 3 Descriptive
statistics of our dataset of
Facebook *pages*

Category	Malicious	Benign
No. of *pages*	627 (31)	1278 (49)
Recent 100 posts	60,306	120,184
Recent 100 posts with URLs	55,233	92,980
Likes (recent 100 posts)[a]	3,447,669	31,680,263
Comments (recent 100 posts)	354,502	1,245,959
Shares (recent 100 posts)	507,964	1,012,151

Numbers in the parentheses indicate verified *pages*
[a]Due to API rate limitations, we had to restrict our data
collection to 50,000 *likes* per post. We had 2 malicious and
291 benign posts which exceeded this limit

Table 4 Number of
malicious posts and *pages* in
each category in our dataset

WOT response	No. of *pages*	No. of posts
Child unsafe	387	10,891
Untrustworthy	317	8057
Questionable	312	8859
Negative	266	5863
Adult content	162	3290
Spam	124	4985
Phishing	39	495
Total	627	20,999

Number of *pages* posting phishing and spam
URLs was the lowest

Table 4 provides a detailed description of the number of posts and *pages* along
with their WOT categories in our dataset of 627 malicious *pages*. We found a
total of 20,999 posts which contained one or more malicious URLs. These posts
engaged a total of 675,162 unique users who *liked*, *commented*, or *shared* these
posts. Interestingly, we found that spam and phishing (two of the most common
types of malicious content studied in literature) were least common in our dataset.
Child unsafe content was the most common, followed by untrustworthy content.

We understand that our sample size of 627 malicious *pages* is not a large dataset
as compared to some of the other studies done on OSNs. However, gathering
Facebook data is a challenging task now. To the best of our knowledge, our dataset of
4.4 million public Facebook posts (from which we identified 627 malicious *pages*)
is one of the biggest samples of Facebook data studied in literature, in terms of
user generated content. The only dataset of Facebook posts (user generated content)
larger than ours was collected by Gao et al. [16]. This dataset was gathered in
2009 by performing large scale crawls on eight regional Facebook networks over
3 months. Authors gathered 187 million posts which originated from roughly 3.5
million users (almost equal to the 3.3 million users + *pages* in our dataset). In
contrast, we gathered all our data through authenticated requests made to the Graph
API over a much larger time span of 16 months. All other studies on user generated
content on Facebook have used much smaller datasets [2, 33, 38]. We discussed
these studies in more detail in Sect. 2.

4 Malicious Pages on Facebook

To understand the differences (and similarities) between malicious and benign *pages*, we studied both the spatial and temporal behavior of these *pages*. We present our findings in detail in this section.

4.1 Spatial Behavior

Most OSNs can be divided into three basic components that make up the social network: *the entity* (user/*page*), *the content* it posts, and its *network* (friends/followers/subscribers). We study all these three components separately.

4.1.1 Entities

We performed term-frequency analysis on unigrams, bigrams, and trigrams obtained from *page* names in our dataset to identify the most prominent entities generating malicious and benign content. Table 5 lists the top 30 unigrams appearing in *page* names in our dataset after removing English stopwords. Manual analysis revealed dominant presence of politically polarized entities and religious groups with keywords like *american, british, english, league, patriot, defense, etc.* in malicious *pages*. Bigram and trigram analysis confirmed wide presence of entities like *British National Party (BNP)*, *The Tea Party*, *English Defense League*, *American Defense League*, *American conservatives*, *Geert Wilders supporters*, etc. We also found some malicious *pages* dedicated to pop bands (One Direction), radio channels (Kiss FM), *pages* using *anonymous* in their names, etc. We manually inspected all the aforementioned *pages* and validated that the *page* names were aligned with the content they published, and were not misleading.

Facebook has been shown to play a significant role in the political context, especially during elections [4, 18, 41]. Researchers have found that knowledge gained by youngsters from Facebook about electoral candidates influenced their evaluation of the candidates [11]. Such impactful role of Facebook on the users prompted us to study and understand the sentiment and emotion of content generated by politically polarized entities in our dataset.

Using the bigram and trigram analysis, we divided *pages* belonging to politically polarized entities into four broad groups based on *page* names to help us study them better. These groups were (1) America (9 *pages*), containing *pages* with "america" or "american" in their *page* name, (2) British National Party (7 *pages*), containing *pages* mentioning British National Party or BNP in their *page* name, (3) Conservative (6 *pages*), containing *pages* with the term "conservative" in the *page* name, and (4) Defence League (11 *pages*), containing *pages* using the phrase "defence league" in the *page* name. We manually verified each *page* to ensure that

Table 5 Word frequency of the top 30 terms appearing in page names in our dataset

Malicious page names				Benign page names			
Keyword	Freq.	Keyword	Freq.	Keyword	Freq.	Keyword	Freq.
news	11	group	5	church	20	county	8
league	11	one	5	center	15	one	8
defense	10	world	5	llc	14	services	8
online	8	videos	5	love	14	fans	8
american	8	national	5	photography	14	south	8
party	8	cricket	5	inc.	13	national	8
english	8	new	4	news	12	life	7
free	7	network	4	united	12	get	7
media	7	bnp	4	school	11	arts	7
truth	7	division	4	team	11	confessions	7
british	6	says	4	community	10	world	7
direction	6	club	4	club	9	health	7
edl	6	tea	4	st.	9	fire	7
forum	5	patriot	4	page	9	dr.	6
radio	5	united	4	cricket	9	beauty	6

We found substantial presence of politically polarized entities among malicious pages

they fit in the group they were assigned. To maintain anonymity, we do not reveal the exact *page* names. We performed linguistic analysis on the content published by these four categories of *pages* separately using LIWC2007 [32]. LIWC is a text analysis software to assess emotional, cognitive, and structural components of text samples using a psychometrically validated internal dictionary. It determines the rate at which certain cognitions and emotions (for example, personal concerns like religion, death, and positive or negative emotions) are present in the text. LIWC has been widely used to study social media content related to politics [37, 39, 40].

We focused our analysis on 12 dimensions in order to profile the linguistic patterns of content published by these groups of *pages*: Positive emotion, negative emotion, anxiety, anger, sadness, money, religion, death, sexuality, past orientation, future orientation, and swear words. Figure 2 shows the results of our analysis. We found high degree of anger in content from all categories. We also observed that only one category of *pages* (British National Party) had more positive emotions than negative emotions. The Defence League *pages* had much higher negative emotions as compared to positive emotions, followed by America *pages*. Conservative *pages* were almost equal in terms of positive and negative emotions. These findings contradicted prior results where researchers found that positive emotions outweighed negative emotions by 2 to 1 for profiles of all German political candidates [40]. We also found substantial presence of content related to religion. These observations are indicative of the kind of influence that politically polarized *pages* in our dataset can have on their audience.

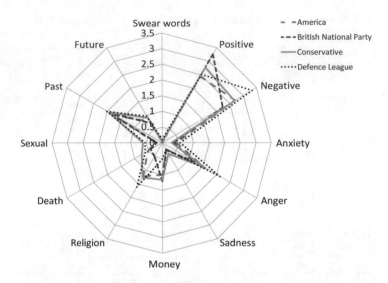

Fig. 2 Linguistic analysis of content produced by politically polarized groups of pages in our dataset. We found high presence of negative emotion, anger, and religion related content

Benign *page* names were found to represent a variety of categories and interests like *photography, school, love, news, confessions*, etc. Bigram and trigram analysis revealed presence of a set of *methodist church pages*. We also found some overlap between malicious and benign *page* names, for example, *One Direction* fan *pages*, and radio channel *pages*. Unlike malicious *pages*, we did not find any fixed category dominating in benign *pages*.

From the above findings, it is evident that politically polarized entities that exist in the real world also have a strong online presence. These results can be used to identify such entities on other social networks as well, and control (if not eliminate) the spread of polar political views online.

4.1.2 Content

Scanning the most recent 100 posts (Sect. 3.3) revealed that almost half of the *pages* (49.28%) in our dataset published 10 or less posts containing a malicious URL. Overall, the median number of domains shared by these *pages* was 24.5. On the contrary, the median number of domains shared by the other half of the *pages* posting more than 10 posts containing a malicious URL (50.72%) was 5. Figure 3 shows the distribution of the number of malicious posts versus the total number of domains shared by all malicious *pages* in our dataset. We found a weak declining trend in the number of domains as the number of malicious posts increased ($r = -0.223$, p-value < 0.01).

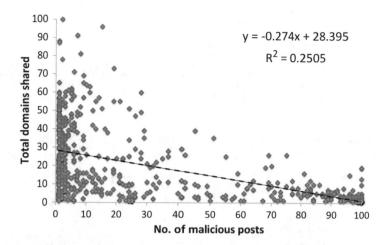

Fig. 3 Number of malicious posts versus all domains published by all 627 *pages* in our dataset. We observed a weak declining trend in the total number of domains when the number of malicious posts published by a *page* increased. Three outliers (sharing 1257, 202, and 140 domains) have been removed in this graph

This declining trend (and negative correlation) indicated that *pages* posting a large number of malicious URLs tend to do so from a small subset of domains. In fact, 84 *pages* in our dataset (13.39%) shared URLs from only one domain. Out of these 84 *pages*, 53 *pages* (8.45%) published more than 90 posts containing a malicious URL, gathering *likes* and *comments* from 55,171 and 31,390 distinct users, respectively. Most certainly, such *pages* exist for the sole purpose of promoting a single (malicious) domain, and are successful in engaging thousands of Facebook users. This sort of activity closely resembles social spam campaigns, which have been studied by multiple researchers [16, 45]. However, since most past research has focused on more obvious threats like unsolicited and targeted spam, advertising, and bulk messaging, other types of malicious content concerned with trustworthiness and child safety have largely remained unaddressed.

Note that there also exist multiple legitimate *pages* on Facebook dedicated to promote a single domain, for example, FIFA World Cup *page* (exclusively posting fifa.com URLs), BBC News *page* (exclusively posting bbc.com URLs), etc. We found 118 *pages* in our benign dataset (9.23%) which were dedicated to promote a particular domain. Such behavior cannot therefore be associated exclusively with malicious activity. Malicious *pages* seem to take advantage of this fact and continue their activity, hiding in plain sight. However, the vocabulary used in the content published by these *pages* can be used to differentiate between the malicious and benign classes using a bag-of-words. We explore this possibility, and report our findings in Sect. 5.2.

Top Domains Table 6 lists the 10 most frequently occurring domains in our dataset of malicious *pages*, along with their WOT classification, Facebook audience, and

Table 6 Top 10 malicious domains in our dataset with their Web of Trust classification, Facebook audience, and Alexa world rank

Domain	WOT class, categories	Posts	Likes I comments I shares	Pages	Page likes	Alexa rank
ammboi.com	Untrustworthy, suspicious, spam, privacy risks	456	666 I 61 I 195	5	109,012	352,191
ridichegratis.com	Untrustworthy	424	428 I 14 I 252	21	2,650,802	–
blesk.cz	Child unsafe, adult content	402	3674 I 2103 I 1494	8	864,554	2924
says.com	Child unsafe	386	387 I 15 I 62	5	97,784	27,684
ghanafilla.net	Untrustworthy, scam, spam, suspicious	296	192 I 8 I 6	3	54,246	1,360,634
9cric.com	Child unsafe	281	1189 I 121 I 177	13	193,348	923,243
perezhilton.com	Child unsafe, adult content	274	26,088 I 3516 I 1128	8	1,701,834	2192
nairaland.com	Untrustworthy, misleading claims or unethical	201	238 I 89 I 31	3	116,131	1329
pulsd.com	Untrustworthy, child unsafe	199	2 I 0 I 0	2	19,020	247,480
970wfla.com	Spam	194	700 I 448 I 280	2	22,486	277,467

Alexa world ranking.[8] For each domain, we calculated the number of posts the domain appeared in, the sum of *likes*, *comments*, and *shares* on all these posts, the number of *pages* the domain appeared in, and the sum of *likes* on all these *pages*. It was interesting to observe that 3 out of the top 10 domains were very famous, and were ranked within the top 3000 domains worldwide on the Alexa ranking. Two of these domains were reported for being unsafe for children and spreading adult content. Although the Internet does not restrict the creation and promotion of adult and child unsafe content, most OSNs including Facebook have established community standards which restrict the display of adult and explicit content [13]. All of the other domains had low Alexa ranking worldwide. Only 3 of the top 10 domains were marked as spam, and none of the domains in the top 10 were reported for phishing or malware, signifying that untrustworthy and child unsafe content is much more prominent on Facebook than traditional forms of malicious content like spam and phishing.

The number of posts and *pages* associated with each of the top 10 domains revealed that there existed multiple Facebook *pages* dedicated to promoting all of these domains. We observed that all of the top 10 domains appeared in 2 or more

[8]http://www.alexa.com/.

pages, and two of the domains appeared in over 10 *pages* (ridichegratis.com in 21 *pages*; 9cric.com in 13 *pages*). At least 4 of the top 10 domains (ammboi.com, ghanafilla.net, pulsd.com, and 970wfla.com) had two or more Facebook *pages* each (3 for ghanafilla.net, 5 for ammboi.com), heavily promoting their respective domains (over 90 out of the 100 posts containing the domain, per *page*). Pages set up for these domains also had a substantial audience, with 6 out of the 10 domains collectively having over 100,000 *likes* on their *pages*. Two of the top 10 domains had over one million *likes* (collectively) on *pages* promoting them. The collective number of *likes*, *comments*, and *shares* on posts was however considerably low as compared to collective *likes* on the *pages*. Only 3 out of the top 10 domains managed 1000 or more *likes* on the posts associated with them. This indicated that while malicious domains in our dataset were successful in gathering a substantial audience in the form of *page likes*, they were not as successful in engaging the audience with their content. We also observed that two of the three domains with high Alexa rank (blesk.cz and perezhilton.com) also had high number of *page likes* and high number of *likes*, *comments*, and *shares* on posts. This signified that domains which were popular (more visited) on the Internet were also more famous on Facebook.

4.1.3 Network

Past research has shown that decentralized networks are prone to *sybil attacks*, wherein malicious entities tend to collude together and attempt to infiltrate the legitimate part of the network [10]. Such attacks have also been studied in context of OSNs [44]. To investigate if such a phenomenon exists for Facebook *pages* too, we analyzed the *like, comment,* and *share* networks for both malicious and benign *pages* in our dataset. Facebook does not provide an API endpoint to gather the list of users who have *liked* (subscribed to) a *page*. However, it is possible to collect the list of users who have *liked, commented* on, or *shared* posts published by a *page*. As described in Sect. 3.3, we collected all *likes, comments,* and *shares* on the most recent 100 posts of all *pages* in our dataset, and analyzed the inter and intra-*page* networks. In particular, we analyzed networks consisting of *pages* and users *liking, commenting on,* or *sharing* posts from two or more *pages* in our dataset (malicious and benign separately) (inter-*page* networks), and networks of *pages* *liking, commenting on,* or *sharing* posts from *pages* within our dataset (malicious and benign separately) (intra-*page* networks). To keep the network size comparable, we averaged out the results for 10 random samples of 627 benign *pages* each (same size as malicious *pages* dataset) drawn from the full 1278 benign *pages* dataset (Fig. 4).

Table 7 shows the details of the network analysis. We found that the Inter-page *likes* network for benign *pages* (83,799 nodes) was much larger and stronger (avg. weighted degree: 41.695) than the Inter-page *likes* network for malicious *pages* (21,947 nodes, avg. weighted degree: 24.273), indicating that a larger number of users *liked* two or more benign *pages* as compared to the number of users who *liked* two or more malicious *pages* in our dataset. More interestingly, we found

Fig. 4 Types of content
published by malicious and
benign pages in our dataset.
Malicious pages published
more links, while benign
pages published more
pictures

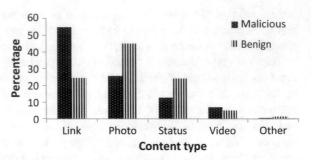

Table 7 Network analysis of *likes*, *comments*, and *shares* networks within and between *pages* in
our dataset

Network type	Total nodes	Total edges	Avg. weighted degree	Density	No. of communities
Malicious (all 627 pages)					
Inter-*page likes* network	21,947	103,683	24.273	0	18
Inter-*page comments* network	3,901	13,957	11.255	0.001	19
Inter-*page shares* network	14,318	67,513	15.796	0	14
Intra-*page likes* network	27	35	8.333	0.05	9
Intra-*page comments* network	9	9	1.667	0.125	3
Intra-*page shares* network	68	65	6.309	0.014	21
Benign (results averaged across 10 random samples of 627 benign pages each)					
Inter-*page likes* network	83,799	390,854	41.695	0	3070
Inter-*page comments* network	2,958	7,722	8.919	0.001	142
Inter-*page shares* network	3,406	10,234	9.920	0.001	30
Intra-*page likes* network	4.3	3.6	0.408	0.075	0.7
Intra-*page comments* network	0	0	0	0	0
Intra-*page shares* network	7.8	6.9	1.168	0.072	1.1

We observed that malicious *pages* had stronger intra-network ties as compared to benign *pages*

stronger ties (avg. weighted degree for Intra-*page* networks) within malicious *pages*
in all aspects (*likes, comments*, and *shares*) as compared to benign *pages*, indicating
collusion and sybil behavior within malicious *pages*. We also found a much larger
number of communities in all Inter-*page* networks for benign *pages* as compared
to Inter-*page* networks for malicious *pages*, indicating a larger and more diverse
audience for benign *pages* as compared to malicious *pages*.

Stronger ties within malicious *pages* prompted us to further investigate the
communities we detected from Intra-*page likes, comments*, and *shares* networks.
Figure 5 shows the network graphs of the detected communities. We observed
that post *sharing* was the most prominent intra-*page* activity, followed by *liking*
and *commenting*. The network graphs also revealed a distinct community of six
Facebook *pages* completely connected to each other in terms of *likes* (Fig. 5a) and
shares (Fig. 5c). Five out of these six *pages* also formed a community in the intra-
page comments graph (Fig. 5b). We manually inspected and observed that all *pages*

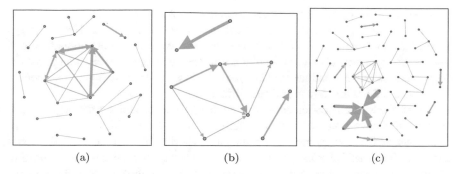

(a)   (b)   (c)

Fig. 5 Network graphs capturing intra-*page* activity of malicious *pages* in our dataset. We found multiple two-node communities and a few bigger communities. (**a**) Malicious *pages liking* each other's posts; intra-page likes network. (**b**) Malicious *pages commenting* on each other's posts; intra-page comments network. (**c**) Malicious *pages sharing* each other's posts; intra-page shares network

in this community belonged to adult stars and promoted pornographic content. This behavior closely resembled a sybil network, and indicated that all these *pages* were controlled by/belong to the same real-world entity (person or organization). We also found multiple two-*page* communities involving politically polarized *pages*, where one *page* heavily engaged in *liking, commenting on*, and *sharing* the other *page*'s content.

4.1.4   Metadata

Analyzing the metadata of posts in our dataset revealed some significant differences in the type of content published by malicious and benign *pages*. Figure 4 shows the distribution of the content type of posts published by all *pages* in our dataset. We observed that more than half of the content published by benign *pages* were photos and videos (50.16%). This percentage went down to 32.42% for malicious *pages*. The metadata also revealed that over half of the posts published by malicious *pages* were links (54.69%), whereas less than a quarter of all posts published by benign *pages* were links (24.45%). These numbers indicate that malicious *pages* are inclined towards posting links, and directing user traffic to external websites. On the other hand, benign *pages* tend to post more pictures, which can be consumed by users without leaving the OSN. In addition to content types, we looked at the status types of posts and found that benign *pages* published almost double the amount of content through mobile devices (23.80%) as compared to malicious *pages* (12.33%).

All *pages* on Facebook have a *category* associated with them, for example *Community, Company, Personal Website*, etc. This category is assigned to the *page* by the *page* administrator(s) at the time of *page* creation, according to the person/organization represented by the *page*, and content that the *page* generates. To see if any subset of categories was more popular among a particular class of *pages* (malicious or benign), we compared category ranks and found strong correlation between category ranks across malicious and benign *pages* (Spearman's $\rho = 0.67$, p—value < 0.01). This indicated that the distribution of malicious and benign *pages* across various categories was fairly similar, and that categories more popular among malicious *pages* were also more popular among benign *pages*. We also compared the *page likes* and *page* mentions (*talking_about_count* field) of malicious and benign *pages*, and did not find any significant differences.

These observations indicated that apart from the type and source of published content, there were no significant differences in the meta information between malicious and benign *pages* in our dataset. Metrics like popularity (*likes*) and user mentions (*talking_about_count*) associated with OSN entities can be used to identify spammers, since they capture the notion of influence of entities in the network [6]. However, similarities in such metrics across malicious and benign *pages* can aid malicious *pages* to continue operating regularly and go undetected for long periods of time, hiding in plain sight.

4.2   Temporal Behavior

We explored the temporal posting activity of all *pages* in our dataset to determine how active the *pages* were, in terms of publishing posts. We also monitored the status of these *pages* daily, for a period of over 1 year to observe any changes in the *pages*' behavior and attributes over time.

4.2.1   Posting Activity

To be able to quantitatively compare the activity of malicious and benign *pages*, we calculated a *daily activity ratio* for each *page*, defined by the ratio of number of days a *page* was active (published one or more posts) versus the total number of days between the first and hundredth post by the *page*.

$$daily\ activity\ ratio = \frac{no.\ of\ days\ active}{no.\ of\ days\ between\ first\ and\ last\ post}$$

Figure 6b shows the *daily activity ratio* plots of all malicious and benign *pages* in our dataset. We observed that 27.43% of all malicious *pages* were active daily as compared to only 8.60% daily active benign *pages*. On average, malicious *pages* were 1.4 times more active daily as compared to benign *pages* in our dataset. We also calculated activity ratio in terms of number of hours and number of weeks, and observed similar results. All activity ratio values were compared using Mann–Whitney U test and the differences were found to be statistically significant (p-value < 0.01 for all experiments) [28]. These differences confirmed that malicious *pages* in our dataset were more active as compared to benign *pages*, and published more frequently.

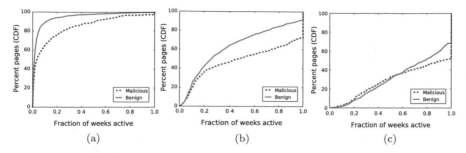

Fig. 6 Daily, hourly, and weekly temporal activity of *pages* in our dataset. We found that malicious *pages* were more active than benign *pages*. (**a**) Hourly activity ratio. (**b**) Daily activity ratio. (**c**) Weekly activity ratio

4.2.2 Attributes over Time

We studied the temporal behavior of all *pages* in our dataset over the period of a year, between October 2015 and October 2016. During this period, we captured daily snapshots of the *page* information for all the *pages* through the Graph API. The aim of this study was to observe the change in attributes of malicious *pages* over time, and to identify if these changes were significantly different from the changes in attributes for benign *pages*. In particular, we studied changes in two types of attributes over time; (a) popularity, and (b) description.

Popularity

To study the change in popularity over time, we computed a *gain factor* corresponding to the change in the number of *likes* on all *pages* in our dataset as follows:

$$\text{gainFactor}_P = \frac{\text{likesOnDayLast}_P - \text{likesOnDayOne}_P}{\text{likesOnDayOne}_P} \times 100$$

where likesOnDayLast_P = no. of *likes* on *page* P on the last day (October 15, 2016), and likesOnDayOne_P = no. of likes on *page* P on the first day (October 16, 2015) of the study. A positive value of the *gain factor* for a *page* P indicates an increase in the number of *likes*, while a negative value depicts a drop in the number of *likes* for P over the span of 1 year.

Figure 7 shows the *gain factor* between malicious and benign *pages* for all *pages* in our dataset. We observed that a larger proportion of malicious *pages* (28.54%) lost *likes* as compared to benign *pages* (20.26%). Contrarily, while computing the average *gain* over all *pages*, we found that malicious *pages* had a larger *gain factor* (32.52%) as compared to benign *pages* (24.03%). This difference, however, was statistically insignificant (p-value > 0.1). Prior statistics show that the average growth rate for a Facebook *page* is 0.64% per week (approximately 33.28% per

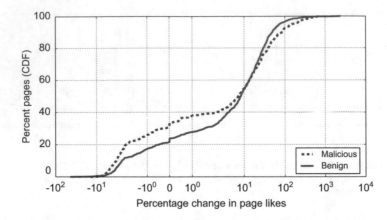

Fig. 7 Percentage change in *page likes* (*gain factor*) over 1 year for all *pages* in our dataset. We observed that a larger proportion of malicious *pages* lost likes over time as compared to benign *pages*

Table 8 Mean values for standard error of estimated gradient and correlation *p*-values for linear model

Metric	Class	Malicious	Benign
Standard error of estimated gradient	μ_{err}	0.8273	0.5583
	σ_{err}	4.6282	4.3492
p-Value for correlation	μ_p	0.016	0.009
	σ_p	0.121	0.082

We obtained low error rates and *p*-values signifying a good fit

year) [25]. Interestingly, this number is much closer to malicious *pages* in our dataset. However, given the statistical insignificance of our results, we cannot conclude that the growth rate of malicious *pages* is more similar to an average Facebook *page* as compared to benign *pages* in our dataset.

We investigated the change in popularity over time in more detail, by computing the rate of change of the number of likes per day for all *pages* in our dataset, to see if there was a statistically significant difference between malicious and benign *pages* with respect to this metric. We modelled the growth rate of *likes* on a *page* as a linear function over time and studied the distribution of the gradient for *page likes* across malicious and benign classes. This technique has been used previously to study popularity on OSNs over time [7]. We observed low values for standard error of the estimated gradient, and significant *p-values* for both classes, signifying good fit (see Table 8).

Figure 8 shows the distribution of the gradients ($\tan^{-1} m$, where $m = \frac{y-c}{x}$; $y =$ no. of likes, $x =$ days, $c =$ intercept) we obtained for malicious and benign classes. We observed that gradients for malicious *pages* were more evenly distributed as compared to benign *pages*. The median gradient value for the malicious class (7.85)

Fig. 8 Distribution of the popularity gradients (in degrees) for malicious and benign pages in our dataset. Although the distributions look different, we did not find the difference to be statistically significant

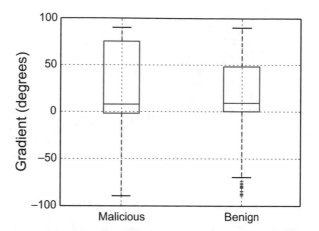

was lower than the median gradient value for the benign class (9.11), but the difference in the two distributions was statistically insignificant ($p-$value $= 0.38$).

Description

Each Facebook *page* has multiple attributes that make up its description, for example, *username, description, general_info, personal_info, category, location, phone_number, mission, bio,* etc. While some attributes (like *username, category*) are available for all *pages*, the presence of other attributes (like *general_info, mission, bio, etc.*) is dependent on the category of the *page*. We examined changes in all such attributes (wherever available) for both, malicious and benign *pages* in our dataset. Figure 9 shows the top 20 attributes in which we observed at least one change during the 1 year period of our study. We came across a total of 44 attributes that were changed once or more.[9] The remaining 24 attributes were changed by less than 1% of all *pages* in our dataset.

We observed a strong correlation between malicious and benign *pages* in terms of the proportion of *pages* changing each attribute, for all 44 attributes ($r = 0.967$, $p-$value < 0.01). This correlation depicted that an attribute that was changed by a large proportion of benign *pages* was changed by a large percentage of malicious *pages* too, and vice versa. For example, the *cover* attribute (cover picture) was changed by 94.44% of benign *pages* and 88.51% of malicious *pages*, while *name* was changed by 6.25% of benign *pages* and 7.49% of malicious *pages*.

We further investigated each of the top 20 attributes individually to see if the changing behavior of any of these attributes could help distinguish between malicious and benign *pages*. We applied the Kolmogorov–Smirnov (KS) 2 sample

---

[9]Exact description for each of these attributes can be found at https://developers.facebook.com/docs/graph-api/reference/page/.

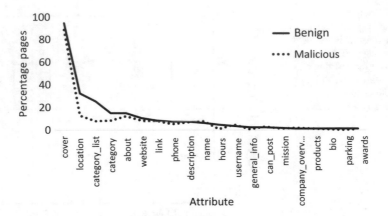

Fig. 9 Top 20 attributes across malicious and benign pages that were changed at least once during 1 year of observation. We identified a total of 44 such attributes, but all remaining attributes underwent one or more changes in less than 1% pages

test to compare the distributions of the number of times each of these attributes were changed by malicious and benign *pages* in our dataset, and found that 19 out of the 20 attributes were not informative (p−value > 0.1). The only statistically significant distribution corresponded to changing behavior of the *category_list* attribute (p−value < 0.05).

The above results corroborate with our previous findings, suggesting minimal presence of a significant difference between malicious and benign *pages*, even in terms of popularity and attribute change behavior over time. These findings further suggest that distinguishing between malicious and benign Facebook *pages* based on spatial characteristics, temporal behavior, and other information associated with these *pages* is a hard and challenging problem. Using all the aforementioned insights and observations obtained from our dataset, we construct a diverse and robust feature set, and attempt to automate the task of identifying malicious *pages* from benign *pages* using supervised learning, as described in the next section.

5   Automatic Detection of Malicious Pages

Past research has shown that URL blacklists and reputation services are ineffective initially, and take time to update [36]. Moreover, lack of blacklists and reputation services for malicious content other than phishing, and malware demand the need for an automated solution to analyze and detect malicious Facebook *pages*. To fulfil this need, we trained multiple supervised learning algorithms on our dataset of malicious *pages* in an attempt to create an effective model for automatic detection of malicious Facebook *pages*, independent of third party reputation services.

Feature Set We extracted a total of 96 features, 55 features from *page* information, and 41 from the posts published by the *pages* in our dataset, to train and evaluate the aforementioned algorithms. Table 9 shows a list of all these features, along with their category and feature type.

Classification Algorithms We experimented with a variety of classification algorithms—Naive Bayesian, Logistic Regression, Decision Trees, Random Forests, and Artificial Neural Networks. We used balanced training and test sets containing equal numbers of positive and negative examples (627 malicious *pages*, and a random sample of 627 benign *pages* picked from the dataset of 1278 benign *pages*), so random guessing results in an accuracy, as well as an area under the receiver operating characteristic (ROC) curve (AUC) of 50%. All models were validated using tenfold cross validation. Although our actual dataset is highly unbalanced, we use a balanced dataset for our experiments in order to obtain a model for better classification of new data, as opposed to a model that would represent our dataset better.

We also performed experiments with unbalanced classes. Note that collecting posts along with their likes, comments, and shares (which we use as features for our analysis) is a time-consuming and computationally expensive task. We therefore performed undersampling on our dataset of malicious *pages* as opposed to collecting more data for benign *pages*. We obtained unbalanced classes in the ratio 1:10 for malicious:benign classes by picking up 10 random samples, each of size 128 (1/10th of the total 1278 benign *pages*) from the malicious class. For each such random sample, we performed tenfold cross validation and averaged out the results across all samples and folds.

In addition, we trained and evaluated bag-of-words models obtained using the textual content present in the posts published by these *pages*. We used the most recent 100 posts published by Facebook *pages* in our dataset for calculating post features and building our bag-of-words. We did not find any explicit distinctive features in our dataset to separate the malicious class from benign, thus making effective automation a hard goal to achieve. We thus tried to build an extensive feature set to capture as much characteristics as possible.

5.1 Supervised Learning with Page and Post Features

Table 10 shows the accuracy and ROC AUC values for various classification algorithms that we applied on the *page* and post level features. We considered post features extracted from the most recent 100 posts generated by the *pages*. A combination of post and *page* level features performed the best, signifying that both the characteristics and posting behavior of *pages* need to be recorded for efficient automatic detection of malicious *pages*. The Logistic Regression classifier achieved highest accuracy of 76.71% with an area under the ROC curve of 0.846.

Table 9 Page and post level features used for training supervised learning models

Category	Feature type	Feature
Page (55)	Boolean (19)	Affiliation, birthday, can post, cover picture, current location, working hours, description present, location, city, street, state, zip, country, latitude, longitude, personal interests, phone number, public transit, website field
	Numeric (34)	Average sentence length for description, average word length for description, parking capacity, category list length, check-ins, no. of email IDs in description, fraction of HTTP URLs in description, description length, fraction of URLs shortened, fraction of URLs active, likes, page name length, no. of subdomains in URLs, path length of URLs, no. of redirects in URLs, no. of parameters in URLs, [no. of !, no. of ?, no. of alphabets, no. of emoticons, no. of English stop words, no. of English words, no. of lowercase characters, no. of uppercase characters, no. of newline characters, no. of words, no. of unique words, no. of sentences, no. of total characters, no. of digits, no. of URLs] in description, description repetition factor, talking-about count, were-here count
	Nominal (2)	Category, description language
Posts (41)	Numeric (41)	Daily activity ratio, audience engaged, [average no. of uppercase characters, average length, average word length, no. of English words, no. of English stop words] for description, message, and name fields, no. of posts containing the field [description, message, name], no. of comments, no. of likes, no. of shares, no. of posts with status_type [added_photos, added_video, created_event, created_note, mobile_status_update, published_story, shared_story, wall post], no. of posts with type [event, link, music, note, offer, photo, video, status], total no. of URLs, total no. of unique domains

The algorithms performed poorly under unbalanced setting. The most probable reason for this poor performance is the lack of data in the malicious class due to undersampling. We therefore preferred models trained on balanced classes.

We performed further experiments by varying the number of most recent posts we considered for generating post features. Figure 10 shows the ROC AUC values achieved by the Logistic Regression classifier with varying post history. We started the experiment by considering 20 most recent posts for post features, and observed an overall increasing trend in performance as we increased the number of most recent posts to 100. We did not go beyond 100 for the post history because our groundtruth dataset for malicious and benign *pages* was derived based on this limit.

The classifier achieved a maximum ROC AUC value of 0.85 (and an accuracy of 77.67%) with a post history size of 80 posts using a combination of *page* and post features. Performance remained unchanged with respect to *page* features, since change in the size of post history does not affect *page* features (and is thus not reported in the figure).

Table 10 Classification accuracy and ROC AUC values for automatically detecting malicious Facebook *pages*

Classifier	Feature set	Balanced		Unbalanced	
		Acc. (%)	ROC AUC	Acc. (%)	ROC AUC
Naive Bayesian	*Page*	63.95	0.685	54.37	0.670
	Post	69.61	0.753	77.29	0.741
	Page + post	70.81	0.776	70.76	0.764
Logistic Regression	*Page*	67.38	0.745	90.81	0.613
	Post	76.55	0.825	90.97	0.740
	Page + post	**76.71**	**0.846**	91.07	0.800
Decision Trees	*Page*	65.55	0.668	90.85	0.493
	Post	71.37	0.720	88.63	0.599
	Page + post	70.81	0.758	90.11	0.660
Random Forest	*Page*	67.86	0.750	90.92	0.745
	Post	74.95	0.829	91.41	0.832
	Page + post	75.27	0.837	91.17	0.856

Logistic Regression classifier performed the best. We used a 1:10 split for malicious:benign classes for unbalanced classes

Fig. 10 ROC area under curve values for Logistic Regression classifier corresponding to different sizes of post history. We observe an overall increase in performance as we increase the number of most recent posts used for computing post features

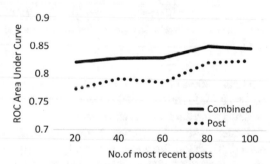

5.2 Supervised Learning with Bag-of-Words

In addition to *page* and post level features, we used a term frequency based bag-of-words model to automatically identify malicious Facebook *pages*. We collected textual content from three sources (wherever present), viz. status message in the post, name and description of the link present in the post (if any).[10] We performed experiments by calculating term frequencies of unigrams, bigrams, and trigrams, and limited our vocabulary size to the top 10,000 features.

A bag-of-words with 10,000 features produced a sparse feature vector. This sparse data prompted us to explore more state-of-the-art learning techniques for fast and effective classification. We chose Sparsenn for this task.[11] Sparsenn is a C

---

[10]https://developers.facebook.com/docs/graph-api/reference/v2.6/post.

[11]http://lowrank.net/nikos//sparsenn/.

Table 11 Classification accuracy and ROC AUC values for automatically detecting malicious Facebook pages using bag-of-words

Classifier	Feature set	Balanced		Unbalanced	
		Acc. (%)	ROC AUC	Acc. (%)	ROC AUC
Naive Bayesian	Unigrams	68.27	0.682	87.10	0.571
	Bigrams	69.06	0.690	88.13	0.611
	Trigrams	69.77	0.697	88.48	0.609
Logistic Regression	Unigrams	74.18	0.795	90.47	0.763
	Bigrams	74.34	0.791	90.47	0.767
	Trigrams	73.93	0.789	90.48	0.782
Decision Trees	Unigrams	68.12	0.678	87.71	0.641
	Bigrams	67.05	0.678	88.41	0.655
	Trigrams	66.63	0.672	88.64	0.658
Random Forest	Unigrams	72.26	0.794	91.64	0.773
	Bigrams	71.80	0.802	91.63	0.778
	Trigrams	72.18	0.794	91.63	0.766
Neural Networks	Unigrams	81.74	0.862	92.41	0.812
	Bigrams	84.12	0.872	92.26	0.734
	Trigrams	**84.13**	**0.900**	92.97	0.826

Artificial neural networks performed the best. We used a 1:10 split for malicious:benign classes for unbalanced classes

implementation of artificial neural networks based on stochastic gradient descent, designed for learning neural networks from high dimensional sparse data. Table 11 presents the results of our experiments.

Under balanced settings, Neural networks (hidden units = 64, and learning rate = 0.07, determined experimentally) on trigrams performed the best, achieving an accuracy of 84.13% with an area under the ROC curve of 0.9. This signified that artificial neural networks trained on the top 10,000 trigrams outperformed all the other learning models including our previous models trained on *page* and post level features (discussed in Sect. 5.1).

We extended our experiments by performing a grid search over post history (number of most recent posts) and bag-of-words size. Using default values for hidden units (16) and learning rate (0.05), we varied the size of the bag-of-words from 1000 through 20,000, and post history from 20 through 100 most recent posts published by the *page*. All these experiments were performed using unigrams, bigrams, and trigrams. Figures 11 and 12 show the varying values of area under the ROC curve for different sizes of post history and bag-of-words respectively. We achieved a maximum ROC AUC value of 0.931 using bigrams with a bag size of 15,000 words and post history size of 100.

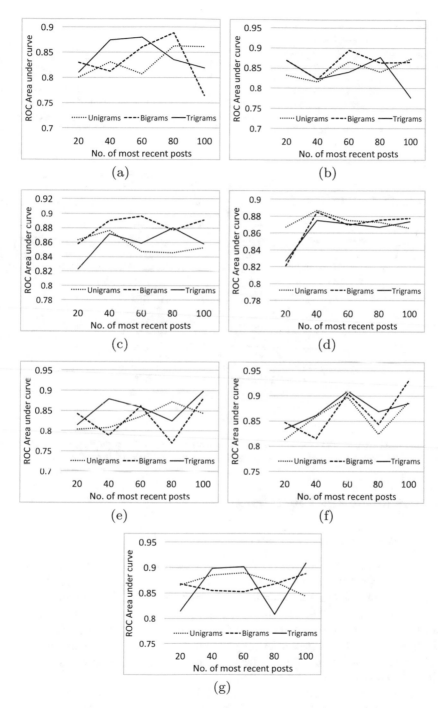

Fig. 11 ROC AUC values obtained by neural networks trained on a bag-of-words for different sizes of bag-of-words. (**a**) 1000 words. (**b**) 2500 words. (**c**) 5000 words. (**d**) 7500 words. (**e**) 10,000 words. (**f**) 15,000 words. (**g**) 20,000 words

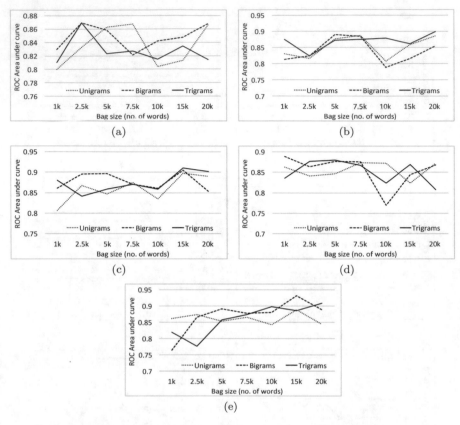

Fig. 12 ROC AUC values obtained by neural networks trained on a bag-of-words for different sizes of post history. (**a**) 20 posts. (**b**) 40 posts. (**c**) 60 posts. (**d**) 80 posts. (**e**) 100 posts

6 Discussion and Limitations

We now discuss the implications of our findings.

Politically Polarized Entities Our analysis revealed the presence of some politically polarized entities in our dataset of malicious *pages*. We verified the actual presence of all such entities in our dataset through manual verification. Interestingly, the presence of such politically polarized entities was obvious despite the fact that none of the events we analyzed had any direct connection with politics, for example, elections. One potential reason for this phenomenon could be a large section of users with polar ideologies marking websites promoting/supporting opposing political views as untrustworthy on Web of Trust.

Entities involved in politics tend to be followed by masses with similar orientation, and is a global phenomenon in the real world. It is likely that such activity exists on online platforms other than Facebook too. We do not propose to debar such activity. However, we believe that extremely polar content should be moderated both online and offline, in order to maintain stability among the masses. An easy way to moderate such entities can be to display nudges or warning messages to users before they subscribe to such *pages* on any online platform [42].

Beyond *Pages* *Pages* on Facebook have a lot in common with Facebook groups and events. Groups and events can also be used to target large audiences at once. Moreover, Facebook has a common definition of "*Page* Spam" for *pages*, groups and events, and explicitly states that *Pages, groups or events that confuse, mislead, surprise or defraud people on Facebook are considered abusive.* Our analysis and results can thus be easily extended to study malicious groups and events as well.

Automatic Identification of Malicious *Pages* Our findings shed some light on subtle differences (like temporal behavior, content type, etc.) between malicious and benign *pages*, which we used to train various supervised learning algorithms to automatically differentiate between malicious and benign *pages*. These findings, however, are based on a limited history (100 posts) of *page* activity. Although it is possible to collect and analyze the entire history for all *pages*, doing so would be time consuming and computationally expensive. Moreover, *pages* can change behavior over time; malicious *pages* may stop spreading malicious content, while benign *pages* may start engaging in posting malicious content over time. To accommodate such changes in behavior, we recommend a self-adaptive model which relies on the most recent activity by the *page*. The history (number of posts) to consider can be decided experimentally. Such a model would be accommodative of the changing behavior of *pages* over time.

Comparison with Past Work Most past work in the space of detecting malicious content on Facebook focuses on individual posts rather than the entities generating these posts (users, *pages*, etc.). This made it nontrivial to compare the performance of our techniques with techniques proposed previously. Further due to the scarcity of publicly available datasets for Facebook data, we were not able to cross evaluate our models. To this end, we intend to anonymize our dataset and make it publicly available for research purposes in the near future.

7 Conclusion and Future Work

In this chapter, we identified and characterized Facebook *pages* posting malicious URLs. We looked beyond traditional types of malicious content like unsolicited bulk messages, spam, phishing, malware, etc., and studied a broader section of content that is deemed as malicious by community standards and *Page* Spam definitions established by Facebook. We focused on Facebook *pages* because of their public nature, vast audience, and inflated malicious activity [8]. Our observations revealed presence of politically polarized entities among malicious *pages*. We also found a substantial number of malicious *pages* dedicated to promote content from a single malicious domain. Further, we observed that malicious *pages* were more active than benign *pages* in terms of hourly, daily, and weekly activity. Network analysis revealed presence of collusive behavior among malicious *pages* that engaged heavily in promoting each other's content. We applied multiple machine learning algorithms on our dataset to automate the detection of malicious *pages*, using *page* and post level features, and bag-of-words. Our experiments showed that artificial neural networks trained on bag-of-words work best in detecting malicious *pages* automatically. We believe that our findings will enable researchers to better understand the landscape of malicious Facebook *pages* that have been hiding in plain sight and promoting malicious content seemingly unperturbed.

In future, we would like to expand our analysis to identify malicious "groups" and "events" on Facebook which largely remain unexplored.

Acknowledgements We would like to thank all the members of Precog Research Group and Cybersecurity Education and Research Centre (CERC) at IIIT Delhi for their constant support and feedback for this work.

References

1. Aggarwal, A., Rajadesingan, A., Kumaraguru, P.: PhishAri: automatic realtime phishing detection on twitter. In: eCrime Researchers Summit (eCrime), 2012, pp. 1–12. IEEE, Piscataway (2012)
2. Ahmed, F., Abulaish, M.: An MCL-based approach for spam profile detection in online social networks. In: IEEE 11th International Conference on Trust, Security and Privacy in Computing and Communications (TrustCom), pp. 602–608. IEEE, Piscataway (2012)
3. Akoglu, L., Chandy, R., Faloutsos, C.: Opinion fraud detection in online reviews by network effects. In: Proceedings of the seventh International AAAI Conference on Weblogs and Social Media, pp. 2–11 (2013)
4. Carlisle, J.E., Patton, R.C.: Is social media changing how we understand political engagement? An analysis of Facebook and the 2008 presidential election. Polit. Res. Q. **66**(4), 883–895 (2013)
5. Castillo, C., Mendoza, M., Poblete, B.: Information credibility on twitter. In: Proceedings of the 20th International Conference on World Wide Web, pp. 675–684. ACM, New York (2011)
6. Cha, M., Haddadi, H., Benevenuto, F., Gummadi, P.K.: Measuring user influence in twitter: the million follower fallacy. In: Proceedings of the Fourth International AAAI Conference on Weblogs and Social Media, pp. 10–17 (2010)
7. De Choudhury, M., Monroy-Hernandez, A., Mark, G.: Narco emotions: affect and desensitization in social media during the Mexican drug war. In: Proceedings of the 32nd Annual ACM Conference on Human Factors in Computing Systems, pp. 3563–3572. ACM, New York (2014)
8. Dewan, P., Kumaraguru, P.: Towards automatic real time identification of malicious posts on Facebook. In: 13th Annual Conference on Privacy, Security and Trust (PST), pp. 85–92. IEEE, Piscataway (2015)
9. Dewan, P., Bagroy, S., Kumaraguru, P.: Hiding in plain sight: characterizing and detecting malicious Facebook pages. In: IEEE/ACM International Conference on Advances in Social Networks Analysis and Mining (ASONAM), pp. 193–196. IEEE, Los Alamitos (2016)
10. Douceur, J.R.: The Sybil attack. In: Peer-to-Peer Systems, pp. 251–260. Springer, Berlin (2002)

11. Douglas, S., Maruyama, M., Semaan, B., Robertson, S.P.: Politics and young adults: the effects of Facebook on candidate evaluation. In: Proceedings of the 15th Annual International Conference on Digital Government Research, pp. 196–204. ACM, New York (2014). http://doi.acm.org/10.1145/2612733.2612754

12. Facebook: what is page spam?. https://www.facebook.com/help/116053525145846 (2015). Accessed 18 Sept 2015

13. Facebook.com: Facebook   community   standards.   https://www.facebook.com/communitystandards (2015). Accessed 12 July 2017

14. Fei, G., Mukherjee, A., Liu, B., Hsu, M., Castellanos, M., Ghosh, R.: Exploiting burstiness in reviews for review spammer detection. In: Proceedings of the Seventh International AAAI Conference on Weblogs and Social Media, pp. 175–184 (2013)

15. Friggeri, A., Adamic, L.A., Eckles, D., Cheng, J.: Rumor cascades. In: Proceedings of the Eighth International AAAI Conference on Weblogs and Social Media (2014)

16. Gao, H., Hu, J., Wilson, C., Li, Z., Chen, Y., Zhao, B.Y.: Detecting and characterizing social spam campaigns. In: Internet Measurement Conference, pp. 35–47. ACM, New York (2010)

17. Gao, H., Chen, Y., Lee, K., Palsetia, D., Choudhary, A.N.: Towards online spam filtering in social networks. In: NDSS (2012)

18. Guardian, T.: Facebook's failure: did fake news and polarized politics get trump elected?. https://www.theguardian.com/technology/2016/nov/10/facebook-fake-news-election-conspiracy-theories (2016). Accessed 12 July 2017

19. Gupta, A., Kumaraguru, P.: Credibility ranking of tweets during high impact events. In: Proceedings of the 1st Workshop on Privacy and Security in Online Social Media, p. 2. ACM, New York (2012)

20. Gupta, M., Zhao, P., Han, J.: Evaluating event credibility on twitter. In: Proceedings of the 2012 SIAM International Conference on Data Mining, pp. 153–164. SIAM, Philadelphia (2012)

21. Gupta, A., Kumaraguru, P., Castillo, C., Meier, P.: TweetCred: real-time credibility assessment of content on twitter. In: Social Informatics, pp. 228–243. Springer, Cham (2014)

22. Jiang, M., Cui, P., Beutel, A., Faloutsos, C., Yang, S.: Catching synchronized behaviors in large networks: a graph mining approach. ACM Trans. Knowl. Discov. Data **10**(4), 35:1–35:27 (2016). Article No. 35

23. Jiang, M., Cui, P., Faloutsos, C.: Suspicious behavior detection: current trends and future directions. IEEE Intell. Syst. **31**, 31–39 (2016)

24. Jindal, N., Liu, B.: Opinion spam and analysis. In: Proceedings of the 2008 International Conference on Web Search and Data Mining, pp. 219–230. ACM, New York (2008)

25. Karma, F.: Study: average growth of Facebook fan pages. http://blog.fanpagekarma.com/2013/03/20/infographic-average-growths-facebook-fan-pages/ (2013). Accessed 12 July 2017

26. Lee, K., Caverlee, J., Webb, S.: Uncovering social spammers: social honeypots+ machine learning. In: Proceedings of the 33rd International ACM SIGIR Conference on Research and Development in Information Retrieval, pp. 435–442. ACM, New York (2010)

27. Lim, E.P., Nguyen, V.A., Jindal, N., Liu, B., Lauw, H.W.: Detecting product review spammers using rating behaviors. In: Proceedings of the 19th ACM International Conference on Information and Knowledge Management, pp. 939–948. ACM, New York (2010)

28. Mann, H.B., Whitney, D.R.: On a test of whether one of two random variables is stochastically larger than the other. Ann. Math. Stat. **18**(1), 50–60 (1947)

29. Mendoza, M., Poblete, B., Castillo, C.: Twitter under crisis: can we trust what we RT? In: Proceedings of the First Workshop on Social Media Analytics, pp. 71–79. ACM, New York (2010)

30. Mukherjee, A., Liu, B., Glance, N.: Spotting fake reviewer groups in consumer reviews. In: Proceedings of the 21st International Conference on World Wide Web, pp. 191–200. ACM, New York (2012)

31. Mukherjee, A., Kumar, A., Liu, B., Wang, J., Hsu, M., Castellanos, M., Ghosh, R.: Spotting opinion spammers using behavioral footprints. In: Proceedings of the 19th ACM SIGKDD International Conference on Knowledge Discovery and Data Mining, pp. 632–640. ACM, New York (2013)

32. Pennebaker, J.W., Chung, C.K., Ireland, M., Gonzales, A., Booth, R.J.: The development and psychometric properties of LIWC2007, Austin, TX: LIWC.net (2007)
33. Rahman, M.S., Huang, T.K., Madhyastha, H.V., Faloutsos, M.: Efficient and scalable socware detection in online social networks. In: USENIX Security Symposium, pp. 663–678 (2012)
34. Ratkiewicz, J., Conover, M., Meiss, M., Gonçalves, B., Patil, S., Flammini, A., Menczer, F.: Truthy: mapping the spread of astroturf in microblog streams. In: Proceedings of the 20th International Conference Companion on World Wide Web, pp. 249–252. ACM, New York (2011)
35. Ratkiewicz, J., Conover, M., Meiss, M.R., Gonçalves, B., Flammini, A., Menczer, F.: Detecting and tracking political abuse in social media. In: Proceedings of the Fifth International AAAI Conference on Weblogs and Social Media, pp. 297–304 (2011)
36. Sheng, S., Wardman, B., Warner, G., Cranor, L., Hong, J., Zhang, C.: An empirical analysis of phishing blacklists. In: Sixth Conference on Email and Anti-Spam (CEAS) (2009)
37. Stieglitz, S., Dang-Xuan, L.: Political communication and influence through microblogging – an empirical analysis of sentiment in twitter messages and retweet behavior. In: 2012 45th Hawaii International Conference on System Science (HICSS), pp. 3500–3509. IEEE, Los Alamitos (2012)
38. Stringhini, G., Kruegel, C., Vigna, G.: Detecting spammers on social networks. In: Proceedings of the 26th Annual Computer Security Applications Conference, pp. 1–9. ACM, New York (2010)
39. Tumasjan, A., Sprenger, T.O., Sandner, P.G., Welpe, I.M.: Election forecasts with twitter: How 140 characters reflect the political landscape. Soc. Sci. Comput. Rev. **29**, 402–418 (2010). https://doi.org/10.1177/0894439310386557
40. Tumasjan, A., Sprenger, T.O., Sandner, P.G., Welpe, I.M.: Predicting elections with twitter: What 140 characters reveal about political sentiment. In: Proceedings of the fourth International AAAI Conference on Weblogs and Social Media, pp. 178–185 (2010)
41. Vitak, J., Zube, P., Smock, A., Carr, C.T., Ellison, N., Lampe, C.: It's complicated: Facebook users' political participation in the 2008 election. CyberPsychol. Behav. Soc. Netw. **14**(3), 107–114 (2011)
42. Wang, Y., Leon, P.G., Scott, K., Chen, X., Acquisti, A., Cranor, L.F.: Privacy nudges for social media: an exploratory facebook study. In: Proceedings of the 22nd International Conference on World Wide Web Companion, pp. 763–770. International World Wide Web Conferences Steering Committee, Republic and Canton of Geneva (2013)
43. WOT: Web of trust api. https://www.mywot.com/en/api (2014). Accessed 12 July 2017
44. Yang, Z., Wilson, C., Wang, X., Gao, T., Zhao, B.Y., Dai, Y.: Uncovering social network Sybils in the wild. ACM Trans. Knowl. Discov. Data **8**(1), 2 (2014)
45. Zhang, X., Zhu, S., Liang, W.: Detecting spam and promoting campaigns in the twitter social network. In: IEEE 12th International Conference on Data Mining (ICDM), pp. 1194–1199. IEEE, Piscataway (2012)

Extraction and Analysis of Dynamic Conversational Networks from TV Series

Xavier Bost, Vincent Labatut, Serigne Gueye, and Georges Linarès

Abstract Identifying and characterizing the dynamics of modern TV series subplots is an open problem. One way is to study the underlying social network of interactions between the characters. Standard dynamic network extraction methods rely on temporal integration, either over the whole considered period, or as a sequence of several time-slices. However, they turn out to be inappropriate in the case of TV series, because the scenes shown on-screen alternatively focus on parallel story lines, and do not necessarily respect a traditional chronology. In this article, we introduce Narrative Smoothing, a novel network extraction method taking advantage of the plot properties to solve some of their limitations. We apply our method to a corpus of three popular series, and compare it to both standard approaches. Narrative smoothing leads to more relevant observations when it comes to the characterization of the protagonists and their relationships, confirming its appropriateness to model the intertwined story lines constituting the plots.

Keywords TV series · Plot analysis · Dynamic social network

1 Introduction

TV series has become increasingly popular these past 10 years. As opposed to classical TV series containing stand-alone episodes with self-contained stories, modern series tend to develop continuous, possibly multiple, story lines spanning several seasons. However, the new season of a series is generally broadcast over a relatively short period: the typical dozen of episodes it contains are usually being aired over a couple of months. In the most extreme case, the whole season is even released at once. Furthermore, modern technologies, like streaming or downloading services, tend to free the viewers from the broadcasting pace, often resulting in an

X. Bost · V. Labatut (✉) · S. Gueye · G. Linarès
Laboratoire Informatique d'Avignon – EA 4128, University of Avignon, Avignon, France
e-mail: xavier.bost@univ-avignon.fr; vincent.labatut@univ-avignon.fr;
serigne.gueye@univ-avignon.fr; georges.linares@univ-avignon.fr

© Springer International Publishing AG, part of Springer Nature 2018   55
M. Kaya et al. (eds.), *Social Network Based Big Data Analysis and Applications*,
Lecture Notes in Social Networks, https://doi.org/10.1007/978-3-319-78196-9_3

even shorter viewing time ("binge-watching"). In summary, *modern* TV series are highly continuous from a narrative point of view but are usually watched in quite a discontinuous way: No sooner was the viewer hooked on the plot than he had to wait for almost 1 year before eventually knowing what came next.

The main effect of this unavoidable waiting period is to make the viewer forget the plot, especially when complex. Since he fails to remember the major events of the previous season, he needs a comprehensive recap before being able to fully appreciate the new season. Such recaps come in various flavors: textual synopsis of the plot sometimes illustrated by keyframes extracted from the video stream; extractive video summaries of the previous season content, such as the "official" recap usually introduced at the beginning of the very first episode of the new season; or even videos of fans reminding, when not commenting, the major narrative events of the previous season. Though quite informative and sometimes enjoyable, such content-oriented summaries of complex plots always rely on a careful human expertise, usually time-consuming. The question is therefore to know how the generation of such summaries can be partially or even fully automated.

To the best of our knowledge, few works in the multimedia processing field have focused on automatically modeling the plot of a movie. In [10], the authors make use of low-level, stylistic features in order to automatically detect the typical three-act narrative structure of Hollywood full-length movies. Nonetheless, such a style-based approach does not provide any insight into the story content and focuses on a fixed narrative structure that generalizes with difficulty to the complex plots of *modern* TV series. The benefits of Social Networks Analysis (SNA) for investigating the plot content of fictional works have recently been emphasized in several articles. Most focus on literary works: dramas [13], novels [1], etc. In the context of multimedia works, SNA-based approaches are even more recent and sparser [9, 17, 18]. However, these works focus either on full-length movies or on stand-alone episodes of classical TV series, where character interactions are often well-structured into stable communities. These approaches consequently do not necessarily translate well when applied to *modern* TV series.

In this paper, we present an SNA-based method aiming at providing some insight into the complex plots of TV series, while solving certain limitations of the previous works. It takes as an input a description of the characters' verbal interactions, and outputs a conversational network. Our method considers not only stand-alone episodes or full-length movies with stable and well-defined communities, but the complex plots of TV series, as they evolve over dozens of episodes. In this case, no prior assumption can be made about a stable, static community structure that would remain unchanged in every episode and that the story would only uncover, and we have to deal with evolving relationships, possibly temporarily linked into dynamic communities. In this case, we are left with building the current state of the relationships upon the story itself, which, by focusing alternatively on different characters in successive scenes, prevents us from monitoring instantaneously the full social network underlying the plot. We thus propose to address this problem by smoothing the sequentiality of the narrative, resulting in an instantaneous monitoring of the current state of any relation at some point of the story.

Our main contributions are the following. In terms of importance, the first is *narrative smoothing*, the method we propose for the extraction of dynamic social networks of characters. The second is the algorithm we introduce to estimate verbal interactions from a sequence of spoken segments. The third is the annotation of a corpus of 109 TV series episodes from three popular TV shows: *Breaking Bad*, *Game of Thrones*, and *House of Cards*. The fourth is a preliminary evaluation of our framework on these data, and a comparison with existing methods. This article extends our paper [6] on several aspects. First, we deepened the background section and the methods description is more detailed, especially the way we estimated verbal interactions between characters. Second, we now present a comprehensive evaluation of the methods that we proposed to handle conversational interactions, and added examples in the application of narrative smoothing to our dataset.

The rest of the article is organized as follows. In Sect. 2, we review in further details the previous works related to SNA-based plot identification. We then describe the method we propose, by first focusing in Sect. 3 on the way the verbal interactions between characters are estimated, before detailing in Sect. 4 the way a dynamic view of the relationships in TV serial plots can be built independently from the narrative pace. In Sect. 5, we first systematically evaluate the algorithm we use for estimating verbal interactions; then, we illustrate how our tool can be used by applying it to the three TV serials of our corpus, and we compare the obtained results to existing methods.

2 Previous Works

In our review, we distinguish between two kinds of works: the first ones consider a static network resulting from the temporal integration over the whole considered period, which we call *complete aggregation*; the second ones extract and study a dynamic network based on a sequence of smaller integration periods called *time-slices*.

2.1 Complete Aggregation

The complete aggregation method consists in building a static network, called a *cumulative network*, in which each node represents a character, and each link models the relationship between the two characters it connects, for the whole considered period of time. It is generally obtained by processing each interaction iteratively, and either updating the weight of the link representing the interaction if it is already present in the graph, or inserting a new link if it is not. Thus, such networks can be weighted. They can also be directed, depending on how the interactions are handled. In the end, a cumulative network is a static graph agglomerating all past relationships, whatever their time ordering. They are widely used in the literature when attempting to apply SNA for analyzing the plot of fictional works.

In [13], the author emphasizes and illustrates the role SNA can play to investigate the plot of literary works. First, after building the network of conversational interactions in a play, the plot, as a sequence of acts occurring over time, is frozen in a spatial, static view that exhibits some underlying patterns: for instance, the conversational network of verbal interactions in Shakespeare's *Hamlet* unveils some critical regions, such as the "region of death" in which the whole tragedy consists. Furthermore, a network-based definition of the protagonists should prevent scholars from applying binary, simplifying categories when considering the main and secondary characters. Finally, the SNA-based notion of community allows to exhibit two distinct spaces in *Hamlet*'s network: a space of legitimacy around Horatio, associated to the modern democratic state, and a space of usurpation around Claudius, related to the old, declining monarchy. The author further illustrates the benefit of SNA for literary studies by considering the question of symmetry in Western and Chinese novels, both at the stylistic level and in the social network of interacting characters.

In [17] and [18], relying on similar observations, the authors make use of SNA to automatically analyze the plot of a movie. The social network of characters (denoted "RoleNet") is built as follows. They first manually characterize the scenes by their boundaries and the characters they involve. They then hypothesize an interaction between two characters whenever they both appear within the same scene. The network is obtained by representing characters as nodes and their interactions by links. These links are weighted according to the number of scenes in which they coappear, resulting in a *cumulative* representation of time. The authors analyze this network through community detection. They apply this approach to so-called "bilateral movies," which involve only two major characters, each of them central in his own community. In [17], the *RoleNet* is used for further investigating the plot, by classifying scenes into one of the two story lines constituting a bilateral movie. In [18], an extended version of the network, without any prior assumption about the number of communities involved, is used as a basis for automatically detecting break points in the story: A narrative break point is assumed if the characters involved in successive scenes are socially distant in the network of characters, as accumulated over the whole story.

In [9], a similar network of interacting speakers is used, among other features, for clustering scenes of two TV series episodes into separate story lines, defined as homogeneous narrative sequences related to major characters. A standard community detection algorithm is applied to the network of speakers, as built upon each episode, before the social similarity between any pair of scenes is computed, as a relevant high-level feature for clustering scenes into substories.

In [15], the authors investigate the character interaction networks of a large database of 173 plays and 580 movie scripts. The interaction between two characters is incremented by 1 if they are both found to speak within a sliding window of 10 successive lines in the screenplays or movie scripts. A number of topological measures such as the average clustering coefficient are processed to describe each network. They are then used as features to perform various classification tasks: distinguishing plays and movies, dating the work, rating it, determining its genre, finding its author, etc.

In summary, cumulative networks can be used as a reliable basis for automatically or manually analyzing the plot of fictional works with well-defined communities, as in plays, full-length movies, or stand-alone episodes of classical TV series.[1]

Nevertheless, for TV serials with complex, evolving, and possibly parallel story lines, such a static approach is not appropriate. Indeed, a cumulative network built over a long period of time, as in TV serials, gets relatively dense and does not enable to extract meaningful information. More specifically, communities in the final agglomerative network undoubtedly always correspond to substories, partially disconnected in the narrative, but the opposite does not generally stand. Some individuals may have been strongly connected to each other at some point of the story, before some of them interact with other people for some time, resulting in a second substory. Once agglomerated in the cumulative network, such changes in the interaction patterns may be obscured. In some extreme cases, distinct narrative sequences may even result in a complete cumulative graph, for instance, in the interaction pattern that follows:

$$s_{12}^{(1)} \dots s_{12}^{(2)} s_{13}^{(3)} \dots s_{13}^{(4)} s_{23}^{(5)} \dots s_{23}^{(6)}$$

where $s_{ij}^{(t)}$ denotes the fact that the ith and jth characters are the only interacting speakers in the tth episode. The three consecutive interaction sequences result in a triangular interaction pattern unable to reflect the three corresponding substories.

2.2   Time-Slices

Some works attempt to take into account the evolution of the social network of the characters when analyzing the plot of fictional works. In [1], the authors emphasize the limitations of the static, cumulative graph when analyzing the centrality of the various characters of the novel *Alice in Wonderland*. A dynamic view of the social network is then introduced, by building successive static views of the network in every chapter, before standard centrality measures are separately computed in each of them and traced over time for some major characters. Each view corresponds to a so-called *time-slice*, or *time-window*.

Though widely used [11] when considering the evolution over time of general networks (i.e., not necessarily narrative ones), time-slice networks, as resulting from the differentiation over some time step of the cumulative network, may still be

---

[1]The website http://moviegalaxies.com/ [12] provides a convenient way of interactively visualizing such cumulative character networks for a database of about 700 movies.

problematic. In [8], the authors focus on the critical issue of the time-slice duration, called "snapshot rate." It must be chosen carefully to allow to capture a sufficient amount of interactions, but not too many, otherwise one may obtain irrelevant network statistics. The authors then describe a way of automatically estimating the natural time-slice for monitoring over time the evolution of a network of daily contacts in a professional context, where the appropriate time-slice is expected to remain constant.

As a smoother alternative to fixed time-windowing, Mutton [14] applies temporal decay to past interactions to estimate the current state of relationships between users of *internet relay chats*. The method is then extended to Shakespeare's plays for monitoring over time the evolution of the network of interacting characters. Nonetheless, we are left with the same kind of issue as with time-windowing approaches: As for the time-slice duration, the ideal value for the decay parameter may be tricky to set.

In order to model the plot of TV serials and allow further analysis, the time-slice should be short enough to capture punctual narrative events related to the social network of characters, but long enough to provide a comprehensive view of the state of the relationships at any point in the story. Unfortunately, getting such a snapshot of the current state of the relationships between the protagonists turns out to be particularly challenging. Unlike the network of physical contacts described in [8], the state of the relationships within a story is not fully monitored at any moment but has to be inferred from the story itself. The narrative usually focuses alternatively on some relationships, possibly belonging to parallel story lines, and only provides a partial view on the network's current state. Some relationships may even take place at the same moment in different places but will be shown sequentially in successive scenes. Figure 1 illustrates the typical sequential nature of the story as being narrated: Three disjoint sets of interacting speakers, possibly at the same time but in different places, are shown sequentially in the story in three successive scenes.

As a consequence, the temporalness of the narrative may be quite different from the temporalness of the underlying network: In particular, the only fact that a group of mutually interacting characters temporarily disappear from the story does not imply that the corresponding characters disappeared from the network. The narrative focusing on these interacting characters may only have been postponed by the filmmaker. Furthermore, the pace of activation of the relationships in the story remains largely unpredictable, especially when multiple, disjoint story lines take place in parallel within the narrative. Figure 2 plots the scene occurrences

Fig. 1   Three different sets of interacting characters from three consecutive scenes

Fig. 2 Narrative frequency
of three character-based story
lines in the first two seasons
of *Game of Thrones*

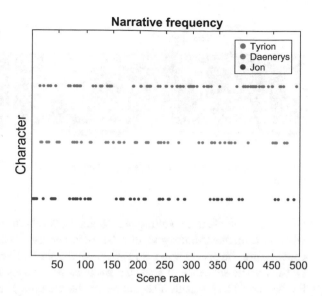

of three major character-based story lines in the first two seasons of *Game of Thrones*. Except in the very beginning of the first season, where Jon Snow and Tyrion Lannister meet each other, the three characters interact within well-separated communities.

As can be seen, the way the story alternatively activates these three major story lines does not seem to follow regular patterns. In such a case, the "ideal" time-slice may be tricky to set. If too large, it will possibly mask the fast changes usually occurring in the most frequently activated story line, for instance, the story centered around Tyrion. If too narrow, it would lead to irrelevant interpretations of the narrative disappearance of some groups of relationships: The narrative disappearance of Jon Snow's story line from scene 400 up to scene 450 does definitely not imply that he does not remain socially active in the meantime in his own community. Therefore, the sequential nature of the story should prevent us from identifying the time of the narrative to the time objectively affecting the social network that the story sequentially unveils.

In the rest of this paper, we introduce a novel way of building the dynamic network of interactions between the characters of TV series that allows to fully capture the instantaneous state of every relationship at any point of the story, whatever the pace of activation of each story line in the narrative.

3   Estimating Verbal Interactions

Getting an accurate view of verbal interactions within TV serials turns out to be quite challenging, either manually or automatically. When considering a sequence of speech turns within a scene, verbal relationships can be stated as soon as a speaker is

Fig. 3 Sequence of speech segments (top) along with corresponding video frames and subtitles (bottom)

talking to an audience, resulting in a directed conversational network, depending on whether someone is talking to, or is being talked by someone. But when a recorded conversation involves more than two speakers, stating who is talking to whom may be tricky. The sequence of speech turns does not convey in itself such information. By just considering such a sequence, as the excerpt of *House of cards* shown on top of Fig. 3, it is impossible to guess who of the three involved speakers is actually speaking to whom.

Such information is usually inferred from complementary sources when available, such as the semantic content of the utterances and/or the video recording of the conversation. In this context, an *utterance* is an uninterrupted spoken segment. In Fig. 3, the images corresponding to the three utterances, along with their linguistic content, help to disambiguate the verbal situation: From who they are looking at when speaking and/or from the use of personal pronouns, we infer that the first speaker (F. Underwood) is clearly speaking to the second one (R. Tusk), while the third one (J. Tusk) is speaking to the first one. However, guessing to whom the second character is specifically talking remains tricky, even from a careful human expertise using this additional information.

In order to avoid such a tedious human expertise, we consider two main options for automatically estimating verbal interactions: the first one classically relies on the co-occurrence of speakers in scenes; the second one is an original contribution and relies on the sequence of utterances within each scene.

3.1 Scene Co-occurrence

Verbal relationships between characters can first be indirectly deduced from their coappearance within semantically homogeneous units. For TV serials, which tend to develop complex story lines over several seasons, each typically consisting of a dozen of episodes, possible units are seasons, episodes, or scenes. Seasons or even episodes turn out to be too wide units to provide an accurate view of the actual verbal interactions within TV serials. Because of parallel story lines, stating as interlocutors all the characters co-occurring in a single season or even episode would result in many irrelevant interactions. Considering the scene as a unit is much wiser: A scene in a movie is defined as a homogeneous sequence of actions occurring at the same place within a continuous period of time. The characters coappearing in a single scene are therefore expected to speak with one another.

A second choice has to be made concerning the kind of character appearance within scenes. Many scenes in movies contain passive characters who do not play any role in the plot and are only physically present but may be talked to by others. Verbal involvement turns out to be much more significant. Although nonverbal relationships, denoted as "observations" in [2], are still possible between main characters, for instance, by only thinking of or by looking at someone else, they usually end up showing verbally in movies. So, by "characters," we will always mean "speakers," and by "occurrence" of the character within a scene, we will mean verbal involvement.

When speech turns, as well as scene boundaries, are explicit, as in plays or movie scripts, the verbal interactions estimated from speaker co-occurrence can be deduced in a fully automatic way. But when the play or the movie is only available as a recorded performance, this information has to be retrieved, either automatically or manually. For TV serials in particular, the scripts are not easily available, or contain only unnormalized and partial information provided on the Web by communities of viewers: We are then left with retrieving speakers as well as scene boundaries.

Though much work has been devoted to automatic detection of scene boundaries, the manual annotation of scene boundaries turns out to be quite straightforward and does not require much time (about 10% of the film real-time duration). The reference scene boundaries, as manually annotated, were thus used for estimating interactions from speakers co-occurrence.

Moreover, automatically detecting "who spoke when" in a movie is quite challenging: Such a task, known in the speech processing field as *speaker diarization* when performed in an unsupervised way, turns out to be especially tricky when applied to TV series, often containing many speakers talking in adverse acoustic conditions (sound effects, background music…). Despite the benefits of multimodal approaches (see, for instance, [4] and [5]), the error rates obtained when applying speaker diarization tools to TV series remain too high (about 50%) to serve as a reliable basis for building interaction networks. As we said, the speakers are thus manually indicated, by labeling the subtitles according to the corresponding speakers. This annotation step is much more time-consuming than for scene delimitation, requiring in average as much time as the real duration of each film.

Nevertheless, though much more relevant than larger-grained units, the scene used as a way of capturing the verbal interactions between characters may result in weak, sometimes irrelevant, interactions: If being at the same place at the same time is usually required to consider that several persons interact, it is rarely sufficient. Figure 4 shows two consecutive dialogs extracted from the TV serial *House of Cards*, and belonging to the same scene. Three speakers are involved, but without

Fig. 4 Two consecutive dialog sequences within the same scene

Fig. 5 Co-occurrence
network based on the
sequence shown in Fig. 4. The
interaction wrongly
introduced is drawn in dash
line

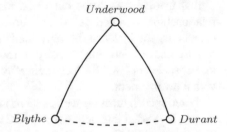

any interaction between the second (*D. Blythe*) and the third (*C. Durant*) ones. The
first speaker (*F. Underwood*) is talking to *D. Blythe* in the first sequence, then is
moving to *C. Durant* and starts discussing with her.

The resulting interaction triangle, based on this scene co-occurrence, is shown
on Fig. 5: *D. Blythe* and *C. Durant* are linked, whereas they are not involved in any
direct verbal interaction. One way of addressing this issue would be to consider
even smaller units than scenes, but such a notion of a "sub-scene" may be confusing
and difficult to define objectively. Another way of facing this problem would be,
instead of globally considering the scene unit, to build the verbal interactions upon
the sequence of utterances in each scene.

We now focus on relationships defined in a *strong sense*, as based on personal
verbal interactions between characters. The resulting network can therefore be
considered as a *conversational network*, in contrast to the co-occurrence network
of characters described in [17, 18] and used in [9].

3.2   Sequential Estimate of Verbal Interactions

Instead of globally considering the scene unit, we choose to tackle this problem by
identifying the verbal interactions from the sequence of speech turns in each scene,
once manually labeled according to the corresponding speakers. In order to estimate
the verbal interactions from the single sequence of utterances, we define and apply
four basic heuristics addressing four possible utterance subsequences. Rules (*1, 3,
4*) apply to subsequences made of three utterances, while Rule (*2*) applies to pairs
of successive utterances.

Rule *(1)*: Surrounded Speech Turn We consider that a speaker s_2 is talking to another speaker s_1 if this s_1 is speaking both after and before s_2, resulting in a speech turns sequence $s_1 s_2 s_1$, where each speech turn is labeled according to the corresponding speaker. Figure 6a shows the subgraph resulting from the application of Rule *(1)* to the speech turns sequence shown in Fig. 4, where each edge is labeled according to the number of times each speaker is considered as talking to another one.

Rule *(2)*: Starting and Ending Speech Turns This rule aims at processing the first and last utterances of each sequence $s_1 s_2 \ldots s_3 s_4$ of speech turns, by adding two links $s_1 \rightarrow s_2$ from the first to the second speaker and $s_4 \rightarrow s_3$ from the fourth to the third one. The network resulting from the application of Rule *(2)* to the sequence of Fig. 4 is shown on Fig. 6b.

The last two rules are introduced to process ambiguous sequences of the type $s_1 s_2 s_3$, where three consecutive speech turns originate in three different speakers: In such cases, the second speaker might be stated as talking to the first one as well as to the third one, or even to both of them. However, such speech turn sequences can often be disambiguated by focusing on speakers involved both before and after the considered sequence.

Rule *(3)*: Local Disambiguation We distinguish two variants of this rule. Rule *(3a)* applies when the second speaker appears before the sequence, but not after, as in $(s_2)s_1 s_2 s_3 (s_4)$. We then consider s_2 is speaking with s_1 rather than with s_3. Symmetrically, Rule *(3b)* concerns the case when the second speaker appears after, but not before the sequence, as in $(s_0)s_1 s_2 s_3 (s_2)$, and is therefore assumed to speak to s_3.

Rule *(4)*: Temporal Proximity When the second speaker is involved in the conversation both before and after the ambiguous sequence, as in $(s_2)s_1 s_2 s_3 (s_2)$, we consider the ambiguous speech turn to be intended for the speaker whose utterance is temporally closer. In the sequence shown in Fig. 4, the fifth, ambiguous utterance would then be hypothesized as intended for the first speaker *D. Blythe*, resulting in the additional link shown in Fig. 6c. The same Rule *(4)* is applied when the speaker s_2 is not involved in the immediate conversational context.

Figures 6 and 7 show the directed interactions identified between any two speakers involved in the scene described by Fig. 4, after jointly applying Rules (1–4). The former considers the rules separately, whereas the latter displays their combined results. On the left-hand part of Fig. 7a, the weight of the links is the number of times one speaker is talking to another one; on its right-hand part (Fig. 7b), the weight is computed as the total duration of the interaction.

We now describe the algorithm we use to build a dynamic network of interacting speakers able to capture the evolution of the narrative content over time.

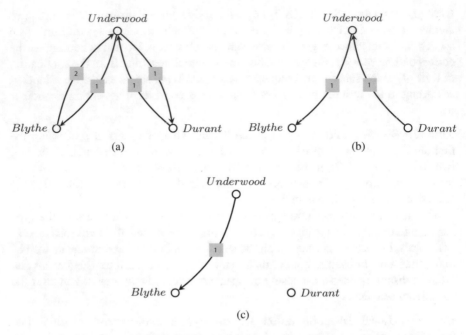

Fig. 6 Number of directed links resulting from the application of Rules (1–4) to the speech turns sequence shown in Fig. 4. (**a**) **Rule** *(1)*. (**b**) **Rule** *(2)*. (**c**) **Rule** *(4)*

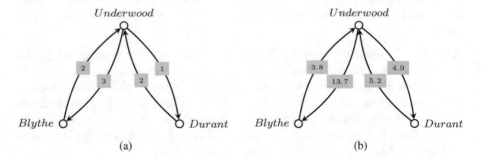

Fig. 7 Directed links resulting from the joint application of Rules (1–4) to the speech turns sequence shown in Fig. 4, with weights corresponding either to the number of interactions (**a**) or to interaction time in seconds (**b**)

4 Dynamic Conversational Network for Plot Modeling

The conversational network is directly extracted from the set of verbal interactions identified at the previous stage. In this section, we first present the method we propose to perform this extraction, then illustrate it with a detailed example.

4.1 Narrative Smoothing

As stated in Sect. 1, we would like to get an instantaneous measurement of the intensity of any relationship at any moment, but from the successive partial views of the underlying network that the narrative provides us. Intuitively, a particular relationship may be considered as especially important at some point of the story if the involved characters both speak frequently and a lot to each other: The time interval needed before the reactivation of the interaction in the narrative is expected to be short, and the interaction time is expected to be long whenever the relationship is active in the plot.

Preliminary Definitions Before putting this idea in practice, we first need to introduce several functions, starting with a measure of the interaction between two speakers during a scene. The rule-based method presented in the previous section allows us to estimate who speaks to whom in a given scene. Based on this, it is possible to compute the total duration of speech one speaker directs at another one during the scene, by summing up the durations of all the utterances identified as such. Let us formally identify the scenes by integer, sequential indices. We obtain $h_{ij}^{(t)}$, the total amount of interaction between two speakers s_i and s_j during the tth scene, by adding the total duration for which s_i talks to s_j, and the total duration for which s_j talks to s_i, during this scene. Note that this value is expressed in seconds, and can be zero if the considered speakers do not speak to each other. Moreover, the function $h_{ij}^{(t)}$ is symmetrical relatively to s_i and s_j, since we make no distinction between the concerned speakers.

The second measure is the *narrative persistence* of the relationship between speakers s_i and s_j at scene t, noted $\Delta_{ij}^{(l)}(t)$. Here, l represents the index of the last scene (relatively to t) during which the considered speakers have verbally interacted. This measure is defined as

$$\Delta_{ij}^{(l)}(t) = h_{ij}^{(l)} - \sum_{t'=l+1}^{t} \sum_{k\neq i,j} \left(h_{ik}^{(t')} + h_{jk}^{(t')} \right) \tag{1}$$

It corresponds to the net balance between the duration $h_{ij}^{(l)}$ of the *last* interaction between the two characters s_i and s_j, and the conversational time (represented by the double sum) that s_i and s_j have devoted separately to other characters s_k since then.

The third measure is the *narrative anticipation*, which is defined symmetrically, and noted $\Delta_{ij}^{(n)}(t)$. Here, n denotes the index of the next scene (relatively to t) during which the speakers will interact. It is defined as

$$\Delta_{ij}^{(n)}(t) = h_{ij}^{(n)} - \sum_{t'=t}^{n-1} \sum_{k\neq i,j} \left(h_{ik}^{(t')} + h_{jk}^{(t')} \right) \tag{2}$$

Quite straightforwardly, it is the difference between the duration of the *next* interaction between the considered speakers, and the time they *will* devote separately to other characters in the meantime.

Now, we can start describing the network extraction process itself. Four possible states have to be considered when monitoring a single relationship over time: (1) the relationship is active in the current scene; (2) it has been active in the story and will be active again later; (3) it was active before, but will no longer be active in the narrative; and (4) it has not yet been active in the narrative.

1. **Relationship currently active in the story.** The first case is the simplest one: Each time the interaction occurs, its intensity can be estimated in a standard way as the duration of the interaction, expressed in seconds. In any scene t where speakers s_i and s_j are hypothesized as talking to each other, the instantaneous weight $w_{ij}^{(t)}$, which represents the intensity of their relationship, is estimated as follows:

$$w_{ij}^{(t)} = h_{ij}^{(t)} \tag{3}$$

where $h_{ij}^{(t)}$ denotes the interaction time, expressed in seconds, between speakers i and j in scene t.

The last three cases are much trickier. Between two consecutive occurrences of the same relationship in the story, it would be tempting to consider that the relationship is still (resp. already) active if it is recent (resp. imminent) enough at each moment considered. This is the method adopted in the time-slice framework described in Sect. 2: As long as the relationship is present in the observation window of the network over time, it is supposed active, and inactive as soon as no longer observed. A smoother alternative based on temporal decay is used in [14].

However, as emphasized in Sect. 2, such a way of handling the past and future occurrences of the relationship is inappropriate for most TV serials. Some interacting characters may be absent from the narrative for an undefined period of time but still be linked in the underlying network, as confirmed by the fact that the last state of the relationship is generally used as a starting point when the characters are reintroduced in the story. Indeed, the temporalness of the narrative should affect a relationship only when at least one of the involved characters interacts with others after and/or before the relationship is active: The relationship between two characters should only get weaker if they interact separately with others before interacting again with one another. This is why, in order to handle the remaining cases, we need to use the previously defined narrative persistence and/or narrative anticipation.

2. **Relationship between two narrative occurrences.** We then define the instantaneous weight $w_{ij}^{(t)}$ of the relationship between the speakers s_i and s_j in any scene t occurring between two consecutive occurrences of their relationship as

$$w_{ij}^{(t)} = \max\left\{\Delta_{ij}^{(l)}(t), \Delta_{ij}^{(n)}(t)\right\} \tag{4}$$

If neither of the two characters speaks to others before they interact again with one another, $w_{ij}^{(t)} = \max\left\{h_{ij}^{(l)}, h_{ij}^{(n)}\right\}$ and the last (resp. next) occurrence of the relation is considered as still (resp. already) fully present in the network, whatever the number of intermediate scenes the narrative introduces in-between to focus on other plot substories.

3. **Relationship after its last narrative occurrence.** The weight of the relationship between the ith and jth speakers in any scene t occurring after its very last occurrence in the narrative is expressed as follows, provided that one of the two characters remains involved in the story by interacting with others:

$$w_{ij}^{(t)} = \Delta_{ij}^{(l)}(t) \tag{5}$$

4. **Relationship before its first narrative occurrence.** Symmetrically, the weight of the relationship between the ith and jth speakers in any scene t occurring before its first occurrence in the story is computed as follows, as long as one of the two characters has already been shown as interacting with other people:

$$w_{ij}^{(t)} = \Delta_{ij}^{(n)}(t) \tag{6}$$

In the very last case, when neither of the two characters is still (resp. already) active, the weight w_{ij} is set to $-\infty$.

Normalization Once the weight $w_{ij}^{(t)}$ has been processed through one of the four possible methods, we use the sigmoid function to normalize it. We note $n_{ij}^{(t)}$ the normalized weight of the relationship between speakers s_i and s_j at scene t:

$$n_{ij}^{(t)} = \frac{1}{1 + e^{-\lambda w_{ij}^{(t)}}} \tag{7}$$

We chose the sigmoid function to perform such normalization for two reasons: (1) to get weights ranging from 0 to 1, and (2) to model the way the past and future states of a relationship in the narrative could influence its current state at some point t. The parameter λ is a parameter of sensitivity to the past and future states of the network and was set to $\lambda = 0.01$ in our experiments (high values imply low dependence on the future and past states). This results in an undirected graph $\mathscr{G}^{(t)}$, capturing the instantaneous state of the social network that the story sequentially unveils.

4.2 Narrative Smoothing Illustrated

Figure 8 shows excerpts of four consecutive scenes in *House of Cards*, involving five individuals. The first two of them, namely, *Francis Underwood* and his wife *Claire*, interact with each other in the first and last scenes (red border), respectively

Fig. 8 Example of application of the weighting scheme to a specific relationship

during 30 and 20 s, whereas Claire interacts in-between 40 s with another person in the second scene (green border) and two other people are talking to one another in the third scene during 50 s.

In the first and fourth scenes, Claire and Francis are interacting with each other: According to Eq. (3), we then set the weights of their relationship to the corresponding interaction times, 30 and 20 s, respectively.

In the second scene, the last interaction between Claire and Francis is on the one hand weakened by the separate interaction of Claire with someone else during 40 s: The resulting *narrative persistence* of the relationship between Francis and Claire then amounts to $\Delta_{12}^{(1)}(2) = 30 - 40 = -10$ (Eq. (1)).

On the other hand, the *narrative anticipation* on the next occurrence of the relationship between Francis and Claire then amounts to $\Delta_{12}^{(4)}(2) = 20 - 0 - 40 = -20$ (Eq. (2)), resulting in an instantaneous weight $w_{12}^{(2)} = \max\{-10, -20\} = -10$ in the second scene.

In the third scene, neither of the two characters is involved: The narrative persistence of their relationship is unchanged, but the narrative anticipation then increases to 20, because no interfering character separates at this point Francis and Claire from their next interaction in the fourth scene. We then have $w_{12}^{(3)} = \max\{-10, 20\} = 20$ and the full resulting sequence of unnormalized, instantaneous weights for the relationship between Claire and Francis is then $(30, -10, 20, 20)$ at the four considered moments.

5 Experiments and Results

In this section, we evaluate the whole framework we introduced for building a reliable and informative dynamic social network of interacting characters in TV serials. We first evaluate the basic heuristics we introduced in Sect. 3.2 to infer the interacting speakers from the sequence of speech turns, once manually annotated according to the corresponding speakers. We then qualitatively evaluate narrative smoothing, our graph extraction method, by comparing it to both types of methods described in Sect. 2. For this purpose, we focus on every TV serial of our corpus, and explore their plots from the dynamics of their underlying social network of characters. We both analyze the obtained networks from the perspective of the protagonists (nodes) and their relationships (links). We now describe the corpus subset we used when evaluating our methods.

5.1   Corpus

Our corpus consists in three very popular TV series: *Breaking Bad* (first three seasons), *Game of Thrones* (first five seasons), and *House of Cards* (first two seasons). We manually annotated the scene boundaries and labeled each subtitle according to the corresponding speaker. The obtained annotations were then used to extract the social networks of characters, by first estimating the verbal interactions according to the rules described in Sect. 3.2 and then by using the existing methods presented in Sect. 2 as well as our own narrative smoothing approach. The resulting networks are publicly available online,[2] along with short videos showing the evolution of the three networks of characters over the seasons considered. Table 1 reports the main features of the first 10 or 11 episodes of each TV series.

As can be seen in Table 1, the duration of the spoken parts in each TV serial covers on average a bit less than half of the total duration of the films. This proportion is expressed as a percentage named *speech coverage* in the table. Speech is more represented in *House of Cards* (coverage $\simeq$ 50% of the total time with 8520 subtitles) than in the other two series.

Speech is uniformly distributed over the scenes, with on average more than 95% of the scenes containing at least one subtitle, which suggests that most social interactions are expressed verbally in these three TV serials.

Furthermore, the average number of speakers by scene remains quite low (ranging from 2.43 to 2.94 depending on the TV series), often resulting in simple patterns of verbal interactions properly handled by applying the basic heuristics described in Sect. 3.2.

We now turn to the evaluation of the accuracy of the four basic rules we use to estimate verbal interactions from the only sequence of speaker-labeled speech turns.

Table 1 Main features of our corpus (first 10/11 episodes)

Corpus	BB	GoT	HoC
Number of episodes	11	10	11
Total duration (H:MM:SS)	8:25:04	8:28:51	8:21:42
Speech coverage (%)	39.5	43.8	50.7
Number of subtitles	6182	6998	8520
Number of speaker occurrences	501	732	951
Number of scenes	206	249	390
Proportion of spoken scenes (%)	94.7	97.2	97.2
Number of speakers per scene (average)	2.43	2.94	2.44
Number of speakers per scene (standard deviation)	1.22	1.52	1.14

---

[2]https://dx.doi.org/10.6084/m9.figshare.2199646.

5.2 Conversational Interactions

The evaluation of the way we estimate verbal interactions from the sequence of speech turns is performed in two ways. First, *directly*, by measuring the performance of our method for achieving the task of estimating interactions. Second, *indirectly*, by measuring the reliability of the cumulative network (cf. Sect. 2.1) resulting from the application of the four basic heuristics introduced in Sect. 3.2. We first describe the episode sample we annotated for this part of the evaluation process.

Sample of Test Episodes In order to evaluate the reliability of the methods introduced for estimating verbal interactions, a subset of test episodes for each of the three TV serials is selected. For each series, the considered subset of episodes is defined so that the distribution of the number of speakers per scene remains representative of the same distribution observed in the first 10 episodes of each series.

The most straightforward way of computing the frequency of the number n of speakers per scene consists in computing the proportion of scenes containing $n = 1, 2, \ldots$ speakers. However, by doing so, one ignores the length of the scenes in terms of utterances, which can be different even between two scenes containing the same numbers of speakers. Using this approach to build our sample could, for instance, result in the same proportion of three-speaker scenes as in the whole corpus, but with significantly shorter scenes, artificially resulting in fewer complex patterns of speech turns, and in better performances when estimating speaker interactions. In order to address this issue, we used a different method, by considering the proportion of *utterances* belonging to scenes with n speakers (instead of the proportion of *scenes*).

The plots constituting the left column in Fig. 9 show the distribution of the number of speakers by scene computed in this way. The lines correspond to the distribution observed in the corpus subset that we introduced in Sect. 5.1, containing about 10 episodes for each of the three TV serials. The points represent the three episode samples we chose as test subsets for each TV serial.

For each of the three resulting TV serial subsets, the same features as those computed in Table 1 are reported in Table 2. For each episode of the three test sets, each utterance is manually labeled according to the speakers it is intended for. For monologues, where no specific listener is targeted, a special *null* label is introduced, and in case of multiple addressees for a single utterance, multiple labels were assigned.

Evaluation Metrics In estimating verbal interaction, the decision is made at the utterance level, by assigning to each utterance the speaker(s) it is intended for. The task then consists in categorizing every utterance among the available speaker classes, with multiple classes allowed if the utterance is labeled as addressed to multiple characters. Standard performance measures used in Information Retrieval in order to evaluate the multi-label categorization task can therefore be used as

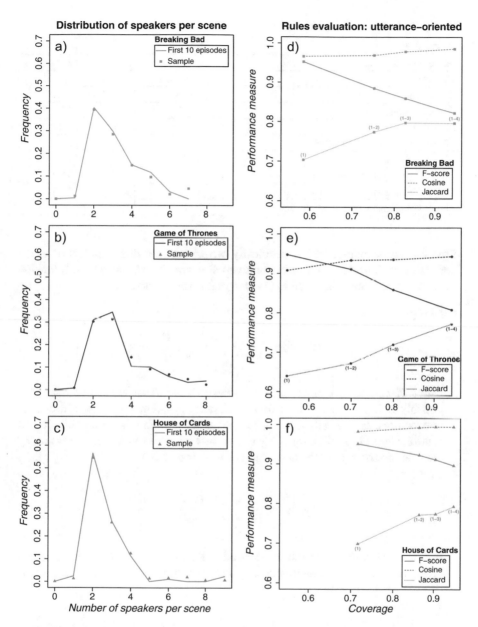

Fig. 9 Distribution of speakers per scene, over the first ten episodes (lines) and over the samples (points) (**a–c**); and step-by-step evaluation of the rules used for sequentially estimating verbal interactions (**d–f**); for the TV series *Breaking Bad* (**a, d**), *Game of Thrones* (**b, e**) and *House of Cards* (**c, f**)

Table 2 Test corpus: main features

TEST subset	BB	GoT	HoC
Number of episodes	4	3	3
Episodes	4, 6, 10, 11	3, 7, 8	1, 7, 11
Total duration (H:MM:SS)	2:59:31	2:37:17	2:29:23
Speech coverage (%)	45.16	48.2	44.2
Number of subtitles	2254	2282	2194
Number of speaker occurrences	202	233	231
Number of scenes	82	81	99
Proportion of spoken scenes (%)	95.1	97.5	97.0
Number of speakers per scene (average)	2.46	2.88	2.33
Number of speakers per scene (standard deviation)	1.14	1.49	1.11

direct evaluation metrics. More specifically, we perform this direct evaluation of the basic rules we introduced for estimating verbal interactions by using the following evaluation procedures discussed in [16] for multi-label categorization:

1. **Recall**:

$$R(\gamma, \mathbb{X}) = \frac{1}{|\mathbb{X}|} \sum_{y \in \mathbb{X}} \frac{|\gamma(y) \cap M(y)|}{|M(y)|} \tag{8}$$

where $\mathbb{X}$ denotes the utterance set; $\gamma(y)$ denotes the set of interlocutor(s), possibly multiple, hypothesized for the utterance y; and $M(y)$ the set of reference interlocutors, as manually labeled, for the utterance y. In this context, the Recall is the average proportion, for every utterance y, of retrieved interlocutors among the reference ones (i.e., the proportion of false negatives).

2. **Precision**:

$$P(\gamma, \mathbb{X}) = \frac{1}{|\mathbb{X}|} \sum_{y \in \mathbb{X}} \frac{|\gamma(y) \cap M(y)|}{|\gamma(y)|} \tag{9}$$

In this context, the Precision corresponds to the average proportion, for every utterance y, of relevant interlocutors among the retrieved ones (i.e., the proportion of false positives).

3. **F-score**:

$$F(\gamma, \mathbb{X}) = \frac{2P(\gamma, \mathbb{X})R(\gamma, \mathbb{X})}{P(\gamma, \mathbb{X}) + R(\gamma, \mathbb{X})} \tag{10}$$

The F-score is the harmonic mean of Precision and Recall, traditionally used in the Information Retrieval domain. Compared to the arithmetic mean, it allows putting more contrast on situations where both the Precision and Recall reach high values.

Besides this direct evaluation of the task of estimating verbal interactions, we also perform an *indirect* evaluation through the assessment of the resulting cumulative conversational network (cf. Sect. 2.1). For this purpose, we compare the cumulative network obtained through our method with a cumulative network extracted manually, which constitutes our ground truth. Following the method described in [2] for a similar purpose, we first convert the adjacency matrix of each one of these two networks into a vector by simple column concatenation. We then measure the similarity between the two resulting vectors. When interactions are weighted, the estimated and ground-truth networks are compared by computing the normalized Euclidean distance and the cosine similarity between the two vectors of edge weights. When focusing on the mere fact that two characters verbally interact with each other whatever the interaction amount, we do not weight interactions, and we evaluate their similarity by computing the Jaccard index of the two sets of edges, both as estimated and as manually labeled. The measures of similarity between the estimated and ground-truth networks are computed both when discarding the first and last utterances of each scene and when considering every speech segment: The first utterance of the next scene is sometimes slightly anticipated at the very end of the current one, possibly resulting in irrelevant interactions.

By using both direct and indirect evaluation metrics, it is possible to measure the performance of the rules we used for sequentially estimating the verbal interactions.

Evaluation Results In evaluating the performance of the rules used for sequential estimate of verbal interactions, we follow a step-by-step process, by successively using in conjunction to the first, the most robust rule, the three remaining ones. The plots on the right side of Fig. 9 show the changes when applying a more and more comprehensive set of rules, from the single first one, denoted *(1)*, to the whole four rules, denoted *(1–4)*: The changes are expressed in terms of coverage (proportion of utterances processed by applying different subsets of rules), and performance (directly through the F-score, and indirectly through the network similarity measures).

Not surprisingly, the more rules are used, the more interactions are hypothesized. As reported in Table 3, the very basic first rule (*surrounded speech turn*) allows on average to hypothesize interlocutors for 62% of the spoken segments. When the whole set of rules is used, decisions are made for 94% of the utterances. The remaining utterances correspond to soliloquies or isolated utterances.

More surprisingly, as can be seen both on the right-hand plots of Fig. 9 and in Table 3, the additional rules *(2–4)* introduce more and more mistakes when hypothesizing interlocutors at the utterance level, resulting in a lower and lower F-score, but in the meantime, the indirect evaluation measures of the resulting network are improving: As can be seen in Table 3, while the F-score decreases on average from 0.95 to 0.84, the Euclidean distance between the estimated and the ground-truth networks decreases from 0.29 to 0.21 if the first and final utterances of every scene are discarded, or from 0.31 to 0.23 if not.

Such a discrepancy suggests that errors made locally when assigning each utterance to the addressed characters do not deteriorate the reliability of the resulting

Table 3 Evaluation of the joint use of the rules applied for sequentially estimating speakers interactions

Rules	Evaluation metrics	TV serial			
		BB	GoT	HoC	Average
(1)	Coverage	0.58	0.55	0.72	0.62
	F-score	0.95	0.95	0.95	0.95
	Precision	0.96	0.95	0.95	0.95
	Recall	0.94	0.94	0.95	0.94
	Jaccard similarity	0.70	0.64	0.70	0.68
	Cosine similarity	0.97–0.96	0.91–0.89	0.99–0.99	0.96–0.95
	L2 distance	0.26–0.29	0.43–0.47	0.19–0.16	0.29–0.31
(1–4)	Coverage	0.94	0.94	0.95	0.94
	F-score	0.82	0.81	0.90	0.84
	Precision	0.84	0.82	0.90	0.85
	Recall	0.80	0.80	0.89	0.83
	Jaccard similarity	0.80	0.77	0.79	0.79
	Cosine similarity	0.99–0.98	0.94–0.93	0.99–0.99	0.97–0.97
	L2 distance	0.17–0.20	0.34–0.37	0.11–0.13	0.21–0.23

The cosine, L2, Jaccard measures are both computed when discarding the first and last segments of each scene, or not

conversational network. Indeed, only a small proportion of the errors made at such a local level (utterance-level *F*-score amounting to zero) introduces irrelevant links in the resulting cumulative network (14.08% for *Breaking Bad*, 13.43% for *Game of Thrones*, and 5.34% for *House of Cards*). Moreover, some errors made at the utterance level by using more and more covering rules allow to retrieve interactions that would otherwise have been missed, or improperly weighted, by carefully applying the only rule *(1)*. The additional rules *(2–4)* tend to introduce correct interactions, but at wrong places, and finally result in more reliable conversational networks, with more actual relationships captured and more representative link weights when measuring interaction intensity. In other words, the errors consisting in misplacing an interaction in time does not affect the cumulative network, in which time is integrated. Though basic, the four heuristics introduced in Sect. 3.2 turn out to be very effective when building such cumulative conversational networks.

We now qualitatively evaluate narrative smoothing, the method we use to build the dynamic conversational network from the interactions between speakers.

5.3   Narrative Smoothing

The Protagonists   We first base our analysis on the protagonists of the considered TV serials, that is, the nodes in the corresponding extracted social networks. We present only a small number of results, which concern characters of particular interest. We characterize them using the *node strength*, a generalization of the concept of *node degree* defined in Graph Theory as the sum of the weights of the links attached to the considered node. In our case, weights are based on spoken interaction durations, so the strength of a character is related to how much and how frequently he speaks to others.

We first focus on Walter White, the main character of *Breaking Bad*, and Tuco Salamanca, one of the drug dealers with whom he is in business. When considering the cumulative network of *Breaking Bad*, that is, the temporal integration over the first 20 episodes, the strength of Walter White (his total interaction time with others) is about 20 times as large as the strength of Tuco: 12,332 s for Walter (rank 1) *vs.* 590 for Tuco (rank 11). By comparison, the left-hand plots of Fig. 10 (plots a–c) display the evolution of their strengths, as a function of time (expressed in terms of scenes). Plots (a) and (b) were obtained through the use of fixed-size observation windows, set to 10 scenes (around half an episode) and 40 scenes (about two episodes), respectively. Plot (c) relies on our narrative smoothing method, so it shows the instantaneous strengths. The plot based on the 40-scene windows (plot b) is consistent with the observation we made on the cumulative graphs, that is, it shows Walter as much more important than Tuco, at any time. It is also the case with the 10-scene windows plot (plot a), but to a lesser extent. In particular, Tuco's strength almost reaches that of Walter around scene 165. But the narrative smoothing plot brings a completely different vision of Tuco's role in the story. From scene 100, Tuco's importance tends to increase and even overcomes the importance of the main protagonist for some time, before suddenly decreasing and reaching almost zero after scene 200. This clearly corresponds to a subplot, or a short narrative episode, ending with Tuco's death, at the end of scene 167 (represented as a vertical line on plots a–c).

We now switch to Daenerys Targaryen and Tyrion Lannister, two major protagonists of *Game of Thrones*. The right-hand plots of Fig. 10 (plots d and e) show how their strengths evolve over the first two seasons of the series, again as a function of the chronologically ordered scenes, and illustrate the limitations of time-windowing approaches. The appearance of Daenerys' story line on-screen has a relatively slow pace in these seasons (Fig. 2) and as can be seen, when the window is too narrow, this creates noisy, irrelevant measurements of her narrative importance (plot d in Fig. 10). It appears very unstable because her story line alternates with many others on the screen. A wider observation window (plot e on the same figure) is more likely to cover successive occurrences of Daenerys in the narrative, but, unlike our narrative smoothing method, prevents us from locating precisely the scenes responsible for Tyrion's current importance. For instance, a local maximum in Tyrion's strength is reached at scene 247 (plot d in Fig. 10), just after a major narrative event took place-the nomination of Tyrion as the King's Counselor (represented as a vertical line in plots d and e). Such an event remains unnoticed when accumulating the interactions during too large time-slices (plot e in Fig. 10), but is well captured by our approach.

Figure 10 also reveals an important property of our way of building the dynamic network. Because the past (resp. future) occurrences of a particular relationship

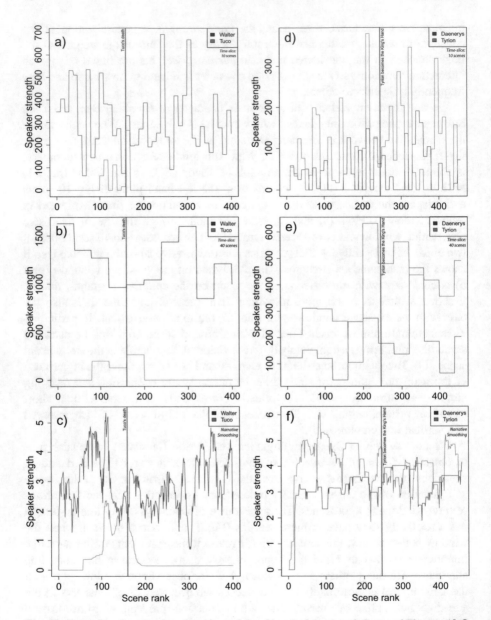

Fig. 10 Strengths of two major characters of *Breaking Bad* (**a–c**) and *Game of Thrones* (**d–f**) plotted as functions of the chronologically ordered scenes, based on 10 scenes (**a, d**) and 40 scenes (**b, e**) time-slices, and narrative smoothing (**c, f**)

are still (resp. already) active as long as the involved characters do no interact with others in the meantime, the respective strengths of the main characters of the story appear remarkably balanced. Although Tyrion looks much more central than Daenerys in the time-slice-based dynamic networks, whatever the size of the observation window, Daenerys is nearly as central as Tyrion in the network based on our narrative smoothing method: few of her acquaintances are shown on-screen as interacting with others. On the opposite, the story focuses more frequently on Tyrion, but also on separate interactions of his usual interlocutors, weakening his instantaneous strength (especially after scene 252): The dynamic strength, as computed after applying narrative smoothing, does not reduce to a global centrality measure, but also corresponds to a more local property, that measures how exclusively a character is related to his/her social neighborhood.

Our results confirm that cumulative networks, by neglecting the temporal dimension, tend to completely miss punctual changes in the importance of certain characters relatively to the plot. The time-slice-based methods can handle the network dynamics; however, our observations illustrate that they cannot properly tackle the narrative issue we described in Sect. 2.2. The choice of an appropriate time window is a particularly sensitive point. By comparison, narrative smoothing captures the state of a relationship at any moment of the plot, using a time scale which directly depends on the narrative pace of the considered series. This allows to finely evaluate the degree of instantaneous involvement of any character in the plot.

The Relationships We now consider relationships between pairs of characters, instead of single individuals. We characterize each relation depending on its weight, that is, the amount of time the characters talked to each other, either cumulated over time-slices, possibly consisting of the whole set of episodes, or smoothed with respect to the narrative. Like for the protagonists, we focus on relationships of particular interest.

Let us consider two relationships in *House of Cards*, representative of two substories: The first one corresponds to a narrative sequence in the story line related to the main character Francis Underwood-his fight with a former ally, the unionist Martin Spinella; the second one is a similar subplot but related to a secondary character, not as frequently present in the narrative, the journalist Lucas Goodwin, who requests the help of the hacker Gavin Orsay to investigate on Francis. Though locally important in these two substories, neither of these relationships lasts long enough to be noticed in the cumulative network, as resulting from the first two seasons of the series: The interaction time amounts to 562 s for the relation between Francis and Martin, and to 294 s for the relation between Gavin and Lucas. These total interaction times remain quite small compared to the central relation between Francis and his wife Claire, amounting to 2319 s.

Nonetheless, once plotted as a function of the chronologically ordered scenes, as shown in the left-hand plots of Fig. 11 (plots a–c), the respective weights of these relationships in the narrative look quite different, whatever the weighting scheme. Both substories, the one based on the relation between Francis and Martin and the

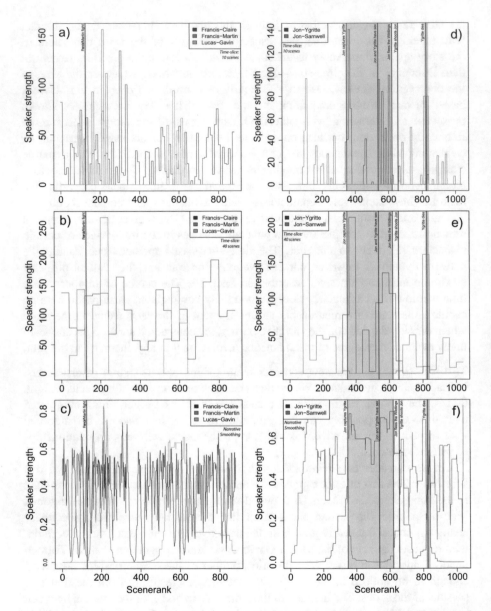

Fig. 11 Weights of several major relationships of *House of Cards* (left column) and *Game of Thrones* (right column) plotted as functions of the chronologically ordered scenes. The first and second rows correspond to 10 and 40 scenes time-slices, respectively, whereas the bottom row shows the results of narrative smoothing

one based on the relation between Lucas and Gavin, turn out to be locally as important as the long-term substory based on the relation between the two main characters Claire and Francis. However, all three ways of monitoring these relationships over time are not equivalent. Agglomerating the interactions within short time-slices (plot a in Fig. 11) makes us miss the continuity of Lucas/Gavin's substory, which occurs *in the narrative* at a slower rate than the substories related to Francis. Conversely, large time-slices (plot b in Fig. 11) allow to capture this substory but agglomerate the two main stages of the relation between Francis/Martin: Before becoming an enemy, Martin is first an ally of Francis. These two parts in the relation correspond to well-separated stages in the narrative, that too large time-slices tend to merge. In contrast, such a break point (materialized in plots a–c by a vertical line located at scene 129) is correctly captured when monitoring the relationship with narrative smoothing.

We now switch back to Game of Thrones, and focus on two links: On the one hand, the romantic relation between Jon Snow and Ygritte, a Wildling, and on the other hand, the friendship between Jon and Samwell Tarly, who also serves in the Night's Watch. Both are represented in the right-hand plots of Fig. 11 (plots d–f), with five important events of the Jon-Ygritte relationship, marked by vertical lines on the plots: (1) their first encounter, when Jon captures Ygritte; (2) the moment when they have sex; (3) their separation, when Jon escapes the Wildlings; (4) the vengeance of Ygritte, when she shoots Jon with her bow and arrows; and (5) the death of Ygritte, during the Wildling attack of Castle Black. The first grayed area represents the duration of the romantic relationship between Jon and Ygritte, and the second is the battle of Castle Black. Samwell and Ygritte belong to two separate social groups, with which Jon alternatively interacts: Samwell when he is at Castle Black with the Night's Watch, and Ygritte when he is north of the wall with the Wildlings. This separation very clearly appears in the plot generated through narrative smoothing (plot f). One can distinguish periods of exclusive relationships: with Sam before the capture of Ygritte (first mark), with Ygritte during their romantic relationship (first grayed area); but also periods where Jon interacts with both, such as the Battle of Castle Black (second grayed area). Our method also allows detecting important events, corresponding to peaks of link weight, such as the first time Ygritte and Jon have sex, or the Battle of Castle Black. By comparison, and as noted before, both methods based on time windows fail to show the continuity of these relationships, which appear to be very sporadic in plots (d) and (e). This is particularly true of the Jon-Ygritte romantic relationship, which appears as completely discontinuous. Of course, this irregularity also hides important events, which do not stand out among these large fluctuations of link weight. Moreover, certain important events are just not associated to important weights, such as the death of Ygritte (rightmost vertical mark).

Our results confirm that cumulative networks are not appropriate for capturing punctual substories supported by specific relationships. Moreover, though much more appropriate for such a task, time-slice approaches suffer from a major drawback: Once fixed, the time slice cannot adapt to the variable rates at which the substories appear in the narrative. By overcoming the narrative contingencies, our narrative smoothing approach allows to monitor more accurately over time any relationship, whatever the way the narrative focuses on it. We could confirm its relevance empirically in a separate study [3, 7]. We generated extractive video summaries based on the networks obtained with narrative smoothing, and evaluated them through a user study during which a large sample of viewers were asked to grade them. The clips produced by hybrid methods combining features extracted from our narratively smoothed conversational networks with lower-level multimedia features obtained the best results.

6   Conclusion and Perspectives

In this paper, we described a novel way of monitoring over time the state of the relationships between characters involved in the usually complex plots of modern TV series. The two methods previously used for this purpose are the cumulative approach, consisting in integrating every relation over the whole considered period of time, and the time-slice approach, consisting in breaking down the time line into smaller discrete chunks. The first one turns out to be relatively inefficient for investigating complex story lines, and a dynamic perspective is more appropriate. The second one complies with this constraint, but defining an appropriate size for the observation window is a very difficult task and constitutes a major drawback: The plots of modern TV series usually consist in parallel story lines shown sequentially on-screen at an unpredictable frequency. As a main consequence, the narrative disappearance in the current scene of some past relationship can usually not be interpreted as a real disappearance, which invalidates the time-slice approach.

To address this issue, we chose to smooth the narrative sequentiality, by considering that the relation between interacting speakers remains active as long as neither of them speaks with others; if so, such separate interactions result in a progressive dissolution of the past link. Symmetrically, the imminence of the next occurrence of the relationship has to increase the link weight.

We then evaluated on our corpus the rules we use for estimating the interacting speakers from the sequence of speaker-labeled speech turns: Though possibly misleading punctually, they result in quite reliable estimates of the characters' relationships. We finally experimentally compared our way of building the dynamic network of interacting speakers, which we call *narrative smoothing*, to both mentioned approaches on the three TV serials of our corpus. Though exploratory and qualitative, our results show that our method leads to more relevant results than both other methods, when it comes to instantaneously monitoring the importance of a particular character or of a specific relationship at some point of the story.

The way some characters temporarily aggregate at some point of the story in a community-like structure suggests that some narrative sequences result in the stabilization, possibly temporarily, of certain areas in the network. By automatically detecting such a narrative stabilization of some groups of relationships, it should be possible to split the whole story into substories, without assuming a static, predefined, community structure. Finally, the statistical properties of such a dynamic network, as based on the smoothing of the narrative, have still to be studied: The relative balance between the important characters suggests, for instance, that the traditional heavy-tailed degree distribution may not stand in this case.

Acknowledgements This work was supported by the French National Research Agency (ANR) GAFES project (ANR-14-CE24-0022) and the Research Federation Agorantic, University of Avignon.

References

1. Agarwal, A., Corvalan, A., Jensen, J., Rambow, O.: Social network analysis of Alice In Wonderland. In: NAACL - Workshop on Computational Linguistics for Literature (2012)
2. Agarwal, A., Kotalwar, A., Rambow, O.: Automatic extraction of social networks from literary text: a case study on Alice in Wonderland. In: International Joint Conference on Natural Language Processing, pp. 1202–1208 (2013)
3. Bost, X.: A storytelling machine? automatic video summarization: the case of TV series. PhD thesis, Université d'Avignon et des Pays de Vaucluse (2016)
4. Bost, X., Linarès, G.: Constrained speaker diarization of tv series based on visual patterns. In: IEEE Spoken Language Technology Workshop, pp. 390–395. IEEE (2014). https://doi.org/10.1109/SLT.2014.7078606
5. Bost, X., Linarès, G., Gueye, S.: Audiovisual speaker diarization of tv series. In: IEEE International Conference on Acoustics, Speech and Signal Processing, pp. 4799–4803. IEEE (2015). https://doi.org/10.1109/ICASSP.2015.7178882
6. Bost, X., Labatut, V., Gueye, S., Linarès, G.: Narrative smoothing: dynamic conversational network for the analysis of TV series plots. In: 2nd International Workshop on Dynamics in Networks (DyNo/ASONAM), San Francisco, pp. 1111–1118 (2016). https://doi.org/10.1109/ASONAM.2016.7752379
7. Bost, X., Gueye, S., Labatut, V., Larson, M., Linarès, G., Malinas, D., Roth, R.: Remembering winter was coming: character-oriented video summaries of TV series (2018)
8. Clauset, A., Eagle, N.: Persistence and periodicity in a dynamic proximity network. arXiv physics.data-an, 1211.7343 (2012)
9. Ercolessi, P., Sénac, C., Bredin, H.: Toward plot de-interlacing in tv series using scenes clustering. In: 10th International Workshop on Content-Based Multimedia Indexing, pp. 1–6 (2012). https://doi.org/10.1109/CBMI.2012.6269836
10. Guha, T., Kumar, N., Narayanan, S.S., Smith, S.L.: Computationally deconstructing movie narratives: an informatics approach. In: IEEE International Conference on Acoustics, Speech and Signal Processing, pp. 2264–2268 (2015). https://doi.org/10.1109/ICASSP.2015.7178374
11. Holme, P., Saramäki, J.: Temporal networks. Phys. Rep. **519**(3), 97–125 (2012). https://doi.org/10.1016/j.physrep.2012.03.001
12. Kaminski, J., Schober, M., Albaladejo, R., Zastupailo, O., Hidalgo, C.: Moviegalaxies - social networks in movies (2012). http://moviegalaxies.com/
13. Moretti, F.: Network Theory, Plot Analysis, vol. 2. Stanford Literary Lab, Stanford (2011)
14. Mutton, P.: Inferring and visualizing social networks on internet relay chat. In: 8th International Conference on Information Visualisation, pp. 35–43 (2004). https://doi.org/10.1109/IV.2004.1320122

15. Suen, C., Kuenzel, L., Gil, S.: Extraction and analysis of character interaction networks from plays and movies. In: Digital Humanities (2013)
16. Tsoumakas, G., Katakis, I.: Multi-label classification: an overview. Department of Informatics, Aristotle University of Thessaloniki (2006)
17. Weng, C.Y., Chu, W.T., Wu, J.L.: Movie analysis based on roles social network. In: ASE/IEEE International Conference on Social Computing and ASE/IEEE International Conference on Privacy, Security, Risk and Trust, pp. 1403–1406 (2007). https://doi.org/10.1109/SocialCom-PASSAT.2012.59
18. Weng, C.Y., Chu, W.T., Wu, J.L.: Rolenet: movie analysis from the perspective of social networks. IEEE Trans. Multimedia **11**(2), 256–271 (2009). https://doi.org/10.1109/TMM.2008.2009684

Diversity and Influence as Key Measures to Assess Candidates for Hiring or Promotion in Academia

Gabriela Jurca, Omar Addam, Jon Rokne, and Reda Alhajj

Abstract Assessing candidates for academic positions or for promotion in academia is a challenging task with many variables to consider. Universities in general and departments in particular may prefer or emphasize diversity, quantity, quality, seniority, juniority, etc. Our case study focuses on the Department of Computer Science at the University of Calgary. Our target is to check how diversity and influence contribute to a department-centric look for hiring or promotion by producing a standard that a candidate may be measured against. We use social network analysis and community detection to measure the influence and diversity of department members. Another measure of diversity could be derived from the number of joint publications between authors and coauthors. The differences in these measures between various positions at the department (including instructors, assistant, associate and full professors) are presented and discussed.

Keywords Bibliometrics · Social network analysis · Community detection · Diversity · Influence

1 Introduction

The hiring or promotion process for candidates in academia is a difficult process where many variables must be considered when assessing a candidate. According to a review done by Long and Fox, science is an institution where universal standards for career attainments are not applied consistently and fairly enough [9]. They compare universalism and particularism with respect to career progress. Universalism argues that scientists should be rewarded only based on their contributions

G. Jurca · O. Addam · J. Rokne · R. Alhajj (✉)
Department of Computer Science, University of Calgary, Calgary, AB, Canada
e-mail: gajurca@ucalgary.ca; rokne@ucalgary.ca; alhajj@ucalgary.ca

© Springer International Publishing AG, part of Springer Nature 2018
M. Kaya et al. (eds.), *Social Network Based Big Data Analysis and Applications*,
Lecture Notes in Social Networks, https://doi.org/10.1007/978-3-319-78196-9_4

to scientific knowledge in the form of publication productivity [12]. Particularism argues that other factors such as race or sex may also affect the selection of candidates for academic positions [9].

Researchers have also debated whether selection committees should examine the publication history of a candidate, or whether committees should place more emphasis on the prestige of the candidate's previous institution [4]. For example, in 1979, Long and Allison conducted a study on the initial academic placement of 239 PhD biochemists and concluded that academic positions are not usually offered based on scientific productivity. Instead, they are offered based on other factors such as prestige of candidate's affiliation [10]. Furthermore, it was also found that previous productivity is a much better predictor of later productivity than doctoral prestige, and that doctoral prestige has no effect on later productivity [10]. In contrast, a critical literature review from 1983 concluded that prestige is strongest correlated with productivity [5]. The best measures of a candidate are not well established, although many researchers and committees strive to make the selection process fairer to reflect reality [12].

In this paper, we shift the focus from the candidate and look toward the department that is interested to make a new hire or a promotion. We provide a methodology to help answer the question of whether a position should be offered based on the existing standards of the department. The goal of the methodology is to provide help in cases where there are too many applicants who must be benchmarked, or when there are too few applicants and it is hard to make a comparison between candidates. We have focused on the Department of Computer Science at the University of Calgary as a case study.

Some researchers showed that the high prestige of a department may be explained by social capital when it cannot be explained by scholarly productivity [4] where this social capital is related to a departments position in a network defined by the exchange of Ph.D. graduates, see [18, 19]. That is, prestige is related to candidates hired from other departments or hired by other departments from the institution being studied. This in turn would produce large interdepartmental networks where influence would flow through the network [4].

Therefore, we assume that a diverse applicant is an attractive candidate for a department that wants to improve its prestige, as the candidate will bring more social capital into the department. We take into account two measures for diversity of a candidate: the number of communities in their egocentric co-publication network and another measure that we propose called influence. While this measure is a more restrictive definition of influence than what is studied in Hayes et al. [6] which considers influence, in general, with respect to brand advertising, advertising on social platforms, etc., it does have the connection with the general influence. The influence measure used in the context of this study takes into account publication productivity and diversity of an author relative to publication productivity and diversity of his/her coauthors.

Using measures of a candidate's publication productivity and diversity, a department may be interested in hiring a candidate who has a larger standing for roles of higher rank, whereas candidates of lower standing may be acceptable for roles of lower rank. The influence measure may also help the department to identify more influential faculty members as outliers who should be promoted to a higher rank or should be rewarded otherwise. Considering diversity and influence will help us to find whether the department can be stratified according to academic roles within the department, namely, full professors, associate professors, assistant professors, and instructors. The influence and diversity measures can then be calculated for a given candidate and compared to the standard of the department for a particular role. Our methodology of classifying a department and generating a standard can also be used by candidates to assess whether a particular department is a suitable fit for them. According to a study, the performance of a department has more effect on the performance of individuals than the other way around [5]. Another study focusses on the random choices of projects in terms of career progress and prestige in terms of citation counts. This view of career progress is not commensurate with the progress metrics used in [17].

The rest of this paper is organized as follows. Section 2 is an overview of related work. Section 3 presents the methodology. Section 4 covers the conducted experiments and the reported results. Section 5 is conclusions and future research.

2 Related Work

There are different bibliometric measures that can be used to rank a candidate for an academic position. These may include a number of publications, citation counts of publications, and impact factors of the journals or venues where publications appeared. However, one common mistake that evaluators make is to use citation counts as evaluation measures, because citation counts take time to accumulate [2] and may be based on criteria other than novelty of the work, for example, criticism of the work. Another mistake that evaluators may make according to Belter is to use the impact factor as a direct measure, because any journal's impact factor is determined by only a small number of publications in that journal (about 10–30%) [2].

A popular measure for the performance of authors is the h-Index [7], which combines the number of publications with a high number of citations for a given author, while being more insensitive to papers with low citation counts. There are also other variations of the h-index. For instance, Bornmann et al. compared nine different variations of the h-index, including m quotient, g index, h(2) index, a index, m index, r index, ar index, and the h_w index [3]. However, neither the h-index nor its variations account for the diversity of a given author. One study has also shown that citation counts may vary across disciplines and must be normalized when considering the h-index [2, 16]. In our case, since we are studying faculty members in the Department of Computer Science, variations in publication counts are assumed comparable across disciplines though they differ in reality from one domain to another even within the same discipline, for example, theory versus applied research.

Some other studies have taken into account the number of coauthors as well. For example, one study looked at normalizing citation counts and paper counts by the number of coauthors, and normalizing citation counts by other papers published in the same journal [15].

Coauthors can also be used to build a co-publication network between a given author and his/her coauthors [14]. Similarly, we use social network in our study to detect the number of communities in each faculty member's egocentric publication network. Another study has also applied social network analysis using social media connections as edges between authors, and then compared the results of the centrality measures (betweenness and closeness) to scientific impact [8]. One limitation of using social media to build a network is that not all authors may be using social media or may be publicizing their social media content.

Other studies have looked at trends in publication performance in terms of the ranks of authoring/coauthoring researchers. For instance, one study found that two factors that currently affect rates of promotion in academic environments are the time spent within a rank and the number of publications within a rank [11]. Another study looked at the performance of newly hired assistant professors across 98 sociology departments [1]. Focusing on the differences between ranks in the Department of Computer Science, we first stratified the Department of Computer Science according to rank (interchangeably role) and then observed the differences between the two measures employed in this study, namely, diversity and influence.

In our work, we have developed a methodology that does not measure the number of publications of a candidate directly, but rather considers diversity of a given candidate's coauthors. Diversity is measured by the number of different communities in an author's egocentric network, and also by the reciprocity between an author and his or her coauthors. The diversity measure of a given author is also affected by how much influential are his/her coauthors.

3   The Proposed Methodology

3.1   The Data Set

The data set examined in this study is related to the faculty members in the Department of Computer Science at the University of Calgary. The list of existing faculty members was accessible online.[1] The same website includes as well the most current role of each faculty member and links to their individual home pages.

The distribution of the faculty members with respect to rank is shown in Table 1. Publications metadata and coauthors of the faculty members were then collected

---

[1]http://www.cpsc.ucalgary.ca/contactus/faculty.

Table 1 The number of faculty members in each rank at the Department of Computer Science at the University of Calgary

Role	Number of individuals
Full professor	15
Associate professor	15
Assistant professor	7
Senior instructor	3
Instructor	3
Total	43

from DBLP, which is an online repository of publications metadata mostly in the field of computer science. The data collection process is further described in Sect. 3.2.

DBLP is a large database with over 2.6 million publications and 1.4 million authors indexed. It was an appropriate choice for this set of individuals because most of their papers are assumed to be published in the computer science domain. Therefore, those papers are assumed to have been indexed by DBLP.

3.2 Data Retrieval

The data for all 43 of the faculty members listed on the department website were manually verified and the metadata of their papers were retrieved during the week of April 10–April 16, 2016. The retrieval of the papers was automated and performed using the DBLP API,[2] where the data was returned in JSON format and parameter h (i.e., the maximum number of search results, also called hits) was set to 10,000.

We did not expect the API to return more than one hit for each faculty member because we had already narrowed our search query for each individual prior to collecting the papers. The only disambiguation was as handled by the software provided by DLBP. The parameter called q, also known as the query string, was retrieved for each name using the API, but manually verified. For example, for Prof. Jon G. Rokne, we searched the DBLP API using his name to retrieve corresponding hits, we browsed the hits, and finally chose the hit with the DBLP ID of ce:author:jon_g_rokne:Jon_G._Rokne based on other forms of cross-checking that we describe in the next paragraph. We then used the verified DBLP ID of each individual to retrieve his/her papers.

To find the most relevant DBLP ID of each faculty member, the most specific name listed on the Computer Science Department page was used in a search query. For example, "Reda S. Alhajj" was used instead of "Reda Alhajj." Also, whenever possible, information from the individual's personal home page was used to narrow down the most representative name for DBLP search. For example, searching "James Tam" resulted in the following hits in DBLP: "James Tam," "Wa

---

[2]http://www.dblp.org/search/api/.

James Tam," "James Tamerius," "James K. Tam," and "K. James Tam." However, upon closer inspection of James Tam's personal home page, we cross-referenced the research papers listed on his home page with those listed for 'James Tam' and concluded that the search term 'James Tam' was appropriate. Our personal familiarity with the department also helped us in navigating the names listed on DBLP and the home pages of individual faculty members.

We also collected the roles of each faculty member, which could be any of the following (listed in the descending order of seniority): full professor, associate professor, assistant professor, senior instructor, or instructor. In one case, for Michael Locasto, we updated his role to Associate Professor based on the information listed on his Web page despite what was listed on the department web page. Indeed, his web page was more current because he had recently been promoted to Associate Professor starting July, 2016.

One individual who we could not retrieve his papers was Michael J. Jacobson Jr., as the DBLP API returned empty results for his name. The DBLP ID for Michael J. Jacobson Jr. provided by DBLP did not return the 50 publications hits that were listed through the DBLP web interface. For further analysis, we considered 17 papers for Michael J. Jacobson Jr. that were found indirectly through coauthorship with the other authors.

After we identified the appropriate DBLP ID for each faculty member, we retrieved all of their papers from DBLP. We also retrieved all of the coauthors of the papers as shown in Fig. 1. We also retrieved all of the papers of each coauthor in order to capture relationships between coauthors who might not involve the author of interest (the authors of interest are all Computer Science faculty members at University of Calgary in our case). We did not retrieve papers published since 2016 because the analysis of the results was initiated in early 2016 and adding more new publications will require repeating all the analysis which would not be feasible.

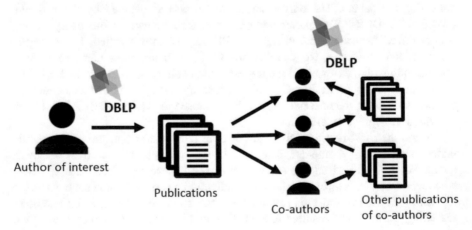

Fig. 1 The different groups of data that were retrieved per author of interest

Follow up research may concentrate on publications beyond 2016 to find out how the concentration of the research since then.

Altogether, the total number of authors retrieved was 2282. So, it was not feasible to manually verify the relevance of each coauthor DBLP ID as we retrieved coauthors' papers. Instead, we constructed our own DBLP IDs based on the names, by replacing the blank spaces and hyphens in the names with "_", replacing the periods with empty strings, transforming all of the letters to lowercase, and finally appending "ce:author:" before each name. For example, "Anna-Maria Brown" may become "ce:author:anna_maria_brown." As a consequence of not verifying the names of the coauthors, there were some coauthors whose papers could not be retrieved due to name ambiguity and misspelling. The average error rate was approximately 10 coauthors per author of interest. The errors were somewhat remedied by the fact that some faculty members had published papers with each other. Therefore, erroneous names of coauthors were sometimes superseded by verified names.

3.3   Influence Values

Influence values were used to assess the impact of authors on coauthors, and vice versa, that is, the impact of particular coauthors on authors. Influence between authors was based on the nature of collaboration, and coauthorship was interpreted as a sign of collaboration between authors.

Influence values take into account the time window of author-author collaboration, as influence is relative to the benefit each author receives during the collaboration period. For example, a professor with many papers published during 2014 does not benefit as much from one paper coauthored by a student, while the student benefits more from the collaboration since the student may have published only one paper in 2014. Therefore, the professor has more influence on the student than the student has on the professor. The influence measure is computed as follows:

$$I_{A_1 A_2} = \frac{\sum_{y=Y_F}^{Y_L} |\text{Papers}_{A_1 \cap A_2}(y)|}{\sum_{y=Y_F}^{Y_L} |\text{Papers}_{A_2}(y)|} \tag{1}$$

The influence ($I_{A_1 A_2}$) of an author A_1 on A_2 is a directional value which is calculated by Formula (1). The first year the two authors have published together is referred to as Y_F, whereas the last year the two authors have published together is referred to as Y_L. The influence is calculated by summing all of the co-publications and dividing that amount by the total number of publications produced by the influenced author (A_2) during the collaboration period. The number of coauthored publications will never be zero, because in such case A_1 and A_2 will not be coauthors. The definition used here is a more restrictive definition that

Fig. 2 The incoming and
outgoing influence calculated
for each author of interest

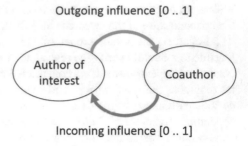

what is studied in Hayes [6] where influence is considered with respect to brands, advertising, and social platforms.

Figure 2 shows that in our case we refer to faculty members in Computer Science at University of Calgary as authors of interest. The influence of the author of interest on a coauthor is referred to as outgoing influence, whereas the influence of the coauthor is referred to as incoming influence.

3.4 Community Detection

Community detection was used to find the diversity of coauthors for an author and to examine the collaboration between coauthors. For example, an author who works with a number of communities might be more sought after by a university, rather than an author who prefers to work within his/her own community. We can also use the community count along with the size of the communities to see whether an author prefers to have one-on-one collaboration with different groups.

To detect communities within an author's network, first we built an undirected egocentric network based on coauthored publications, where coauthors also had edges with other coauthors. The weight of an edge represents the number of papers that had been published between the two connected authors. Community detection was done by iteratively calculating edge betweenness of all edges and then removing edges with the highest betweenness, as described by Girvan and Newman in 2004 [13]. As edges with the highest betweenness are removed, modularity of the graph is calculated and then used to partition the dendrogram to give the highest modularity, as described in [13].

The egocentric network was built in *R* using the igraph package, and edge betweenness was calculated using edge.betweenness.community function of the igraph package. The communities were found by calculating modularity in a way similar to the implementation described in the following tutorial: http://www.sixhat.net/finding-communities-in-networks-with-r-and-igraph.html.

Similar to Influence calculation, we considered the time frame of the collaboration as we detected the communities. That is, we computed communities for every single year of an author of interest's career. That is, if Bob has published papers from 1994 to 2015, we first detect communities from his egocentric network which is derived from papers published in 1994. Then, we detect the communities from his egocentric network which is derived from papers which were published in 1994 and 1995, followed by community detection from his egocentric network derived from papers published in 1994, 1995, and 1996. We repeat the community detection process until we reach his final year, 2015. The number of communities detected each year can be used to see whether Bob has been part of many communities as time passed, or if he has been part of few communities over the period of his academic career.

4 Experiments, Results, and Discussion

4.1 Influence

The incoming and outgoing influence of each faculty member was calculated and then averaged across different roles. Figure 3 shows the average outgoing influence of the different roles in the Department of Computer Science at University of Calgary. The instructors and senior instructors were included under one group. We decided to combine them because the combined group size is only six individuals in total. Assistant professors have the smallest average outgoing influence value of 0.47, while full professors have the largest average outgoing influence value of 0.64 (see Fig. 3).

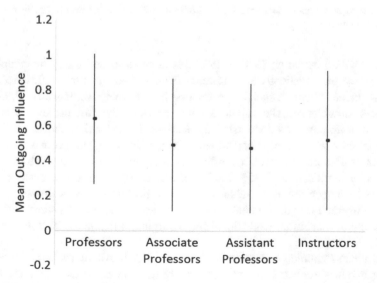

Fig. 3 The mean outgoing influence across different roles. The error bars represent population standard deviation

Table 2 Statistical information for the mean incoming influence displayed in Fig. 4 and the mean outgoing influence displayed in Fig. 3

Role	N	Mean $I_{Outgoing}$	SD	Mean $I_{Incoming}$	SD
Professors	1308	0.64	0.37	0.11	0.13
Associate professors	812	0.49	0.38	0.20	0.20
Assistant professors	278	0.47	0.37	0.28	0.30
Instructors	53	0.52	0.40	0.41	0.30

SD represents standard deviation, and N represents the total number of coauthors of the individuals in the role group

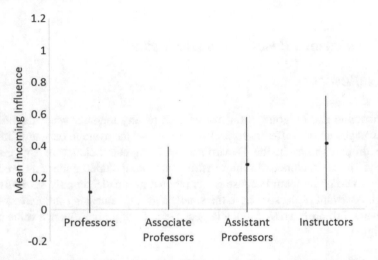

Fig. 4 The mean incoming influence across different roles. The error bars represent population standard deviation

The standard deviation (SD) of the roles appears to be largely overlapping in Fig. 3. However, statistically the difference between the mean of the full professors and the mean of the associate professors is extremely statistically significant with 95% confidence. The unpaired t test between the full professors and the associate professors was done using the statistical information shown in Table 2. An unpaired t test was also performed between the associate professors and the assistant professors. However, the difference between the two mean values was not statistically significant with 95% confidence. Similarly, the t test revealed a lack of difference between the mean values of assistant professors and instructors. Therefore, the average outgoing influence of associate professors, assistant professors, and instructors is similar, while the average outgoing influence of full professors is larger.

The mean incoming influence is shown in Fig. 4. All of the mean incoming influence values are less than the mean outgoing influence values, with the largest mean being 0.41 for the instructors and the smallest being 0.13 for the full professors. According to the unpaired t test, the difference between full professor

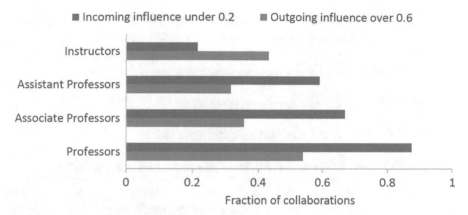

Fig. 5 The fraction of collaborations for each role that either contained an outgoing influence over 0.6, or an incoming influence under 0.2

mean and associate professor mean, the difference between associate professor mean and assistant professor mean, and the difference between assistant professor mean and instructors mean are all extremely statistically significant with 95% confidence. Therefore, the roles of the faculty members are correlated with the average incoming influence values. However, in all categories there are outliers who either have a larger outgoing influence or a smaller outgoing influence than the average. These upper-tier good performing fellows (i.e., outliers) are cases who the department may want to consider for promotion or rewarding.

We then examined distributions of more specific influence values. In Fig. 5, the fraction of outgoing collaborations that contain an influence score greater than 0.6 are displayed. In Fig. 5, the fraction of incoming collaborations that contain an influence score less than 0.2 are also shown. The values of 0.2 and 0.6 were chosen based on the mean values of full professors in Figs. 3 and 4 because we wanted to compare the other roles to the higher standards of full professors.

Figure 5 shows that the fraction of collaborations that contain an outgoing influence score greater than 0.6 increase as the academic rank of the role increases from assistant professor to full professor. Curiously, instructors have a larger proportion of outgoing influence scores over 0.6 than assistant professors and associate professors. One issue lies in the nature of the data set, as there are only six instructors in our data set (including three senior instructors). Some of the instructors have a very small number of publications (1 or 2 were retrieved), and some of those publications are single-author papers. However, some of the instructors have a number of publications comparable to an assistant or associate professor. The data perhaps shows that instructors have a flexible role in the department, where they can choose to be more academically involved, or they can focus on other aspects such as teaching. Nonetheless, some of the instructors show a large outgoing influence on those they work with.

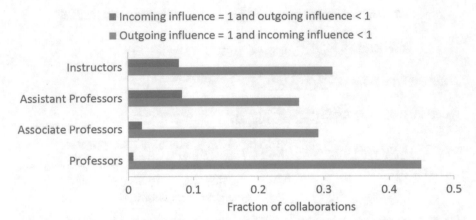

Fig. 6 The fraction of collaborations for each role that either contained an incoming influence of 1 or an outgoing influence less than 1. The fraction that contains an outgoing influence of 1, but an incoming influence of less than 1 is also shown

On the other hand, as the academic rank increases from instructor to full professor, the fraction of incoming influence that scores less than 0.2 increases. That is, as the academic rank of an individual in the department increases, the amount of influence that the person receives through coauthorship is diminished. Figure 5 shows that approximately 90% of full professor collaborations resulted in an incoming influence less than 0.2. One explanation for the diminished incoming influence is the fact that professors are more likely to coauthor many papers with many other people.

To look at some more extreme cases, we examined the fractions of collaborations that resulted in an influence of 1 for only one of the parties involved (Fig. 6). In Fig. 6, we see the large disparity between providing full influence and receiving full influence in the full professorial role. Again, an increase in the fraction of full outgoing influence is seen from the assistant professorial role to the full professorial role, but the instructor role appears to have a large frequency of full outgoing influence. The number of full incoming influence decreases as academic rank increases.

4.2 Communities

Edge-betweenness community detection was performed in order to see the variety of groups that each faculty member is working with. A hiring committee may want to examine the diversity of a candidate and how the candidate's level of diversity fits into the department, and in what role. In order to assess community detection, the algorithm was performed on the most current network of Dr. Reda

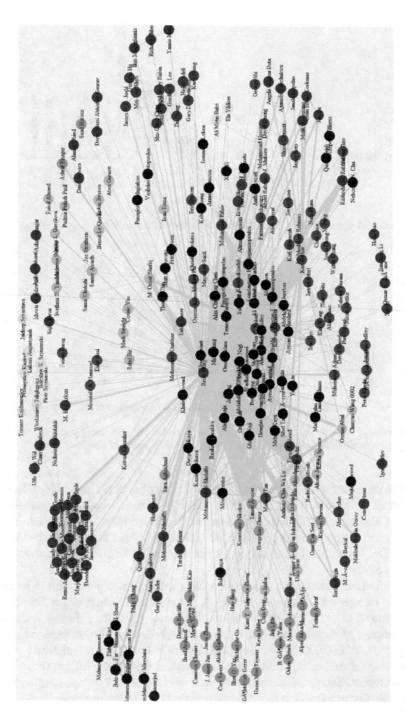

Fig. 7 Communities of Dr. Reda Alhajj's egocentric network of coauthorship

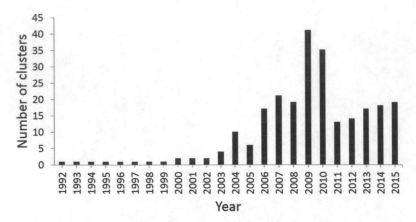

Fig. 8 Number of communities in Dr. Reda Alhajj's egocentric network with respect to time

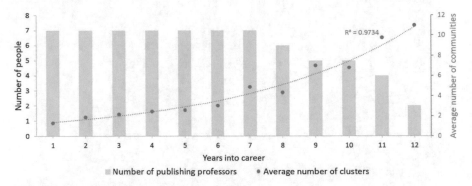

Fig. 9 Number of assistant professors with respect to the number of publishing years. An average number of communities within the egocentric networks of individuals are shown

Alhajj, the professor with the highest number of coauthors (255) in the Department of Computer Science at the University of Calgary. Figure 7 shows the egocentric network of Dr. Reda Alhajj and the coloring of the related individual communities. At first glance, the algorithm appears to have reasonably detected the communities, although Dr. Reda Alhajj has not yet evaluated the results of the community detection process.

The community detection process was then applied on a yearly basis to the cumulative coauthorship matrices. In Fig. 8, the number of clusters for Dr. Reda Alhajj is shown to span the years he has been actively publishing. Figure 8 shows that the number of clusters have fluctuated for the professor, with the peak number of clusters as 49 in 2009. Since the networks are constructed cumulatively each year, the case for decreasing cluster numbers may not be that the full professor has left communities. The decreases in the number of clusters may be due to groups of coauthors that have merged to form a larger group in the following year.

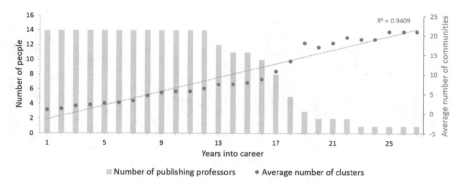

Fig. 10 Number of associate professors with respect to the number of publishing years. An average number of communities within the egocentric networks of individuals are shown

Community detection was then performed on yearly cumulative networks of every other faculty member. The yearly number of clusters was counted for each faculty member. The results for the assistant professors are shown in Fig. 9, where we have also aligned the publishing years of each person. For example, the figure shows that all seven associate professors have been publishing for at least seven years, and there exists two associate professors who have been publishing for 12 years. Aligning the publishing years of individuals allows us to see the average number of clusters that the associate professors had when they started publishing and through the years after that. A hiring committee may want to compare the number of clusters of a candidate to a department average while also considering the number of years the candidate has been publishing.

The average number of clusters increases as the publishing years increase for associate professors (see Fig. 9). The increase in the average number of clusters has also been fitted with an exponential curve where the R^2 value is 0.9734. The maximum number of clusters reached for associate professors is 12.

The average number of clusters for associate professors is shown in Fig. 10. All the 14 associate professors have been publishing for at least 12 years, which shows that a greater extent of experience is required for this role. Note that Associate Professor Michael J. Jacobson Jr. has not been included in Fig. 10 because his papers were not successfully retrieved through the DBLP API. The maximum average number of clusters is 20, which is reached approximately after the 18th year of publishing. Interestingly, the increase in the average number of clusters has been fitted to a linear trend line with an R^2 value of 0.9409. A linear trend line fitted to the average number of clusters of the assistant professors resulted in an R^2 value of 0.8879. Therefore, the rate of increase in the average number of clusters slightly diminishes from assistant professors to associate professors.

Finally, the average number of clusters for full professors is shown in Fig. 11. All of the full professors have been publishing for at least 17 years. The average number of clusters increases steadily, if not exponentially, until about the 17th year of publishing. After 17 years, the average number of clusters began to fluctuate

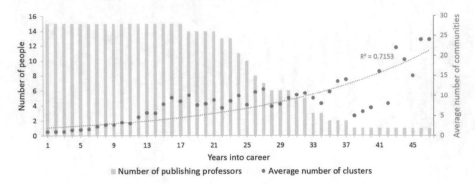

Fig. 11 Number of professors with respect to the number of publishing years. An average number of communities within the egocentric networks of individuals are shown

year by year, but without much of an increase. After the 32nd year of publishing, three full professors remain and the average number of clusters fluctuates with larger differences than before. Since there are a smaller number of professors in the department with more than 32 years of publishing, the cluster averages were not stabilized after that.

A literature review done in 1983 that examined other studies on the effect of age on productivity found that a decline or flattening of productivity for researchers older than 30–40 years has been reported in literature [5]. Although we are examining the publication years, presuming that faculty members started their research careers in their early 20s, faculty members with 17 years of publication history may be approximately 40 years of age. Surprisingly, as a whole, the average number of clusters increases as publishing years increase beyond 37 years. There is only one full professor with more than 37 years of publishing. Results of the data series for the average number of clusters were fitted with an exponential curve, but with a poorer R^2 value of 0.7153.

A comparison between the associate professors (Fig. 11) and the full professors (Fig. 11) reveals an interesting difference between the groups. For associate professors, the maximum average number of clusters was 21 which was reached after 21 years of publishing, yet the full professors have reached only an average of 8 clusters after 21 years of publishing. The discrepancy may be due to generational differences between the two groups, as the full professors may have reached 21 years as the mark for years before the associate professors.

Perhaps the introduction of different technologies throughout the years has made it easier for younger authors to collaborate and communicate more easily with different groups of coauthors. A 1983 literature review of other bibliometric studies also found that cohort differences have been reported with respect to publication productivity, where newer cohorts had less of a decline in publication productivity [5]. For example, considering the number of publishing years of a candidate may be important while assessing the number of different communities in the candidate's network.

The average number of communities per year was not examined for instructors, as there was a low number of instructors with publications (4).

4.3 Limitations

One of the limitations of our study was that other online repositories of computer science publications could have been used to complement the data available from DBLP. For instance, other databases related to computer science publications include Association for Computing Machinery (ACM) digital library, IEEE XPLORE, SCOPUS, etc. Furthermore, we realized that some researchers publish across multiple domains. Thus, in the future we may have to collect publication metadata from diverse sources. For example, some computer scientists may publish bioinformatics papers in cancer journals, and as a result those publications are not listed in DBLP.

Another limitation of our study was that names were not disambiguated for the retrieved coauthors. As a result, the number of clusters in the networks may have been overestimated, as there may be more connections between coauthors than what has been retrieved. Also, single authored papers were not considered while calculating the influence score for each author, and as a result, the influence score may have been skewed. Single authored papers should be considered because authors with single authored papers will have a smaller incoming influence than what has been currently calculated.

5 Conclusions and Future Research

In this study, we investigated how a hiring committee might assess the fitness of an academic candidate into a department. Our methodology may also be used by a candidate to assess whether the department could be a suitable fit. First, we considered the needs and interests that a department may have to consider while searching for a hire or promotion. Our case study was the Department of Computer Science at the University of Calgary, where we were able to demonstrate effect of the two measures described in this paper, namely, influence and diversity.

First, we used an estimation of influence to find differences between different roles in the department, namely, assistant professors, associate professors, full professors, and instructors. Differences in average influence were found between the different roles that can be used to categorize candidates. It is assumed that new hire or promotion should at least meet the minimum requirements for the role he or she is applying for within the department. Second, to find the minimum standard of influence and diversity, we used edge-betweenness community detection to find the number of communities that each existing faculty member had in their egocentric network. The results were plotted for different roles in the department. We found that each role (or rank) in the department varied with respect to the minimum number of years during which members had published, and that the rate of change in the number of communities slightly varied by role. The fitted trend lines for the average number of communities in the egocentric networks per year may be used to assess a candidate for a particular role.

As future work we are interested in addressing the limitations that were stated in Sect. 4.3. Also, the methodology may be applied to data sets from other departments or other universities in order to find whether there exist a common standard for different roles in a department, in terms of influence or diversity of communities in an egocentric network. Also, it may be interesting to consider the affiliations of the coauthors, as a diverse amount of affiliations among the coauthors may indicate influence and diversity of the author of interest. We may also want to consider the number of coauthors when calculating the influence between two authors. For example, two coauthors may have a larger influence on each other in a two-author paper as compared to an eight-author paper where the two appear in the list without knowing their real contribution or influence.

References

1. Bauldry, S.: Trends in the research productivity of newly hired assistant professors at research departments from 2007 to 2012. Am. Sociol. **44**(3), 282–291 (2013)
2. Belter, C.W.: Bibliometric indicators: opportunities and limits. J. Med. Libr. Assoc. **103**(4), 219 (2015)
3. Bornmann, L., Mutz, R., Daniel, H.D.: Are there better indices for evaluation purposes than the h index? A comparison of nine different variants of the h index using data from biomedicine. J. Am. Soc. Inf. Sci. Technol. **59**(5), 830–837 (2008)
4. Burris, V.: The academic caste system: prestige hierarchies in PhD exchange networks. Am. Sociol. Rev. **69**(2), 239–264 (2004)
5. Fox, M.F.: Publication productivity among scientists: a critical review. Soc. Stud. Sci. **13**(2), 285–305 (1983)
6. Hayes, J.L., King, K.W., Ramirez, A. Jr.: Brands, friends, & viral advertising: a social exchange perspective on the ad referral processes. J. Int. Mark. **36**, 31–45 (2016)
7. Hirsch, J.E.: An index to quantify an individual's scientific research output. Proc. Natl. Acad. Sci. U. S. A. **102**(46), 16569–16572 (2005)
8. Hoffmann, C.P., Lutz, C., Meckel, M.: Impact factor 2.0: applying social network analysis to scientific impact assessment. In: 2014 47th Hawaii International Conference on System Sciences (HICSS), pp. 1576–1585. IEEE, New York (2014)
9. Long, J.S., Fox, M.F.: Scientific careers: universalism and particularism. Annu. Rev. Sociol. **21**, 45–71 (1995)
10. Long, J.S., Allison, P.D., McGinnis, R.: Entrance into the academic career. Am. Sociol. Rev. **44**, 816–830 (1979)
11. Long, J.S., Allison, P.D., McGinnis, R.: Rank advancement in academic careers: sex differences and the effects of productivity. Am. Sociol. Rev. **58**, 703–722 (1993)
12. Merton, R.K.: A note on science and democracy. J. Legal. Pol. Soc. **1**, 115 (1942)
13. Newman, M.E., Girvan, M.: Finding and evaluating community structure in networks. Phys. Rev. E **69**(2), 026113 (2004)
14. Otte, E., Rousseau, R.: Social network analysis: a powerful strategy, also for the information sciences. J. Inf. Sci. **28**(6), 441–453 (2002)

15. Petersen, A.M., Wang, F., Stanley, H.E.: Methods for measuring the citations and productivity of scientists across time and discipline. Phys. Rev. E **81**(3), 036114 (2010)
16. Radicchi, F., Fortunato, S., Castellano, C.: Universality of citation distributions: toward an objective measure of scientific impact. Proc. Natl. Acad. Sci. U. S. A. **105**(45), 17268–17272 (2008)
17. Sinatra, R., Wang, D., Deville, P., Song, C., Barabási, A.L.: Quantifying the evolution of individual scientific impact. Science **354**, aaf5239-1–aaf5239-8 (2016)
18. Subbian, K., Aggarval, C., Srivasta, J.: Content-centric flow mining for influence analysis in social streams. In: Proceedings of the 22nd ACM International Conference on Information & Knowledge Management (2013)
19. Subbian, K., Aggarval, C., Srivasta, J.: Mining influencers using information flows in social streams. ACM Trans. Knowl. Discov. Data **10**, 26:1–26:28 (2016)

Timelines of Prostate Cancer Biomarkers

Gabriela Jurca, Omar Addam, Jon Rokne, and Reda Alhajj

Abstract Prostate cancer is a deadly disease which affects men, yet there are few specific and sensitive biomarkers that can be used to diagnose and provide prognosis for the disease. Prostate specific antigen (PSA) is a serum protein that is one of the most well-established biomarkers in prostate cancer, but it lacks specificity and sensitivity. As a result, researchers are in search of other biomarkers such as genes or proteins that can be used for prostate cancer diagnostic test. In order to save effort, qualitative reviews based on primary studies are manually performed to characterize genes and proteins as emerging biomarkers. However, one problem is that less effective biomarkers might not be explicitly addressed as poor biomarkers in primary studies due to publication bias. We use text mining to provide a tool to examine whether biomarkers are emerging or decreasing in terms of publication popularity. In addition, we provide a tool to examine the increasing or decreasing popularity of gene families with respect to prostate cancer research. Selected biomarkers which have been labelled as emerging in qualitative reviews are evaluated using our approach. We also provide public access to our web portal to allow users to explore genes or gene families that they are interested in.

Keywords Prostate cancer · Text mining · Data mining

1 Introduction

Prostate cancer is a deadly disease, and it is also heterogeneous genetically. There are different molecular types of prostate cancer, which may respond differently to therapy. Therefore, it is important to find sensitive and specific biomarkers for prostate cancer. The most well-established biomarker for diagnosis and prognosis of prostate cancer is the prostate specific antigen (PSA), which is a serum protein

G. Jurca · O. Addam · J. Rokne · R. Alhajj (✉)
Department of Computer Science, University of Calgary, Calgary, AB, Canada
e-mail: gajurca@ucalgary.ca; rokne@ucalgary.ca; alhajj@ucalgary.ca

© Springer International Publishing AG, part of Springer Nature 2018   105
M. Kaya et al. (eds.), *Social Network Based Big Data Analysis and Applications*,
Lecture Notes in Social Networks, https://doi.org/10.1007/978-3-319-78196-9_5

that was first identified in the 1970s [16]. However, the PSA test has also been cited as a cause of overdiagnosis, due to its lack of specificity and sensitivity [3, 5, 13]. One problem with the PSA test is that PSA levels may be elevated in the absence of prostate cancer, such as in the case of benign prostate hyperplasia (BPH), which is a noncancerous expansion of the prostate.

Due to the problems with the PSA test, researchers are searching for other biomarkers that can indicate prostate cancer, categorize prostate cancer, or be used as targets for treatment. In order to study prostate cancer biomarkers, researchers may look at previous research done in order to find promising or emerging biomarkers. However, with approximately 130,000 publications on prostate cancer in the PubMed database, it can be difficult for researchers to track down emerging prostate cancer biomarkers. It is also worth considering publication bias, where studies with negative results on biomarkers may not be published. That is, some biomarkers published in older articles may appear to be important in spite of the fact that no authors have published positive results on those biomarkers recently. One reason might be that authors may not be publishing papers on some of the biomarkers because experiments may have found the biomarkers to be ineffective or no more sensitive or specific than PSA. It is important that researchers know which biomarkers are emerging, or whether the biomarkers have been decreasing in research popularity, so that they may focus on the most effective biomarkers and bring the biomarkers to clinical use faster.

In this paper, we find the status of a gene or gene family by indicating whether it is attracting increased research interest (emerging) or decreasing interest (demerging) assessed by the number of papers publishing results on the gene or the gene family. In doing so, we are tacitly assuming that an increase in publication frequency on the subject of the usefulness of a particular biomarker for prostate cancer indication indicates an increasing awareness of this biomarkers utility and appropriateness for diagnosis. We use the Moving Linear Regression Angle (MLRA) algorithm [17] to assess the shape of the data allowing an easier detection of local extrema and trends in the time series of papers mentioning a specific biomarker.

Some genes emerged for some time but recently they have started demerging but not quite enough to be considered as demerging. Therefore, we use fuzzy membership to define the emerging and demerging memberships. We propose a membership function that can be used to detect if the gene is emerging or demerging. The membership function uses the MLRA generated time series and a weighted score for the year.

The rest of the paper is organized as follows: Sect. 2 discusses some of the background that is related to our work. Section 3 explains our methodology for fetching and analyzing the data. Section 4 lists the experiments and interprets the results. Finally, Sect. 5 concludes our paper and presents our future work.

2 Related Work

The related work that we examined comprised of reviews that listed current and emerging biomarkers in prostate cancer. There are different types of prostate cancer biomarkers, such as the types listed by Martin et al. [6]: biomarkers found in the serum, the tissue, and also molecular signatures. Cancer biomarkers can be used to calculate predisposition to prostate cancer, diagnosis, and prognosis. For example, specific biomarkers can be used to find the aggressiveness of the prostate cancer of a patient, or to find whether the prostate cancer will metastasize or not. The reviews that we examined mentioned genes and proteins other than PSA protein (KLK3 gene) that could be used as biomarkers for prostate cancer. Selected biomarkers identified as "emerging" by various reviews are listed in Table 1. Most of the mentioned biomarkers in Table 1 are standalone genes, but in one case there is also a gene signature, such as the gene fusion between the gene TMPRSS2 and either gene ERG or ETV1. Gene fusions may occur as a result of chromosomal translocations.

Table 1 Biomarkers identified as current or emerging by related work

Biomarker gene symbol	Related work (# references that authors used)
KLK2	Parekh et al. in 2007 [9](3),
	Sardana et al. in 2008 [12](4),
	Reed et al. in 2010 [11](4),
	Velonas et al. in 2013 [14](3)
TGF-β 1	Sardana et al. in 2008 [12](3),
	Velonas et al. in 2013 [14](5)
EZH2	Sardana et al. in 2008 [12](2),
	Martin et al. in 2012 [6](6),
IL6	Sardana et al. in 2008 [12](3),
	Martin et al. in 2012 [6](3),
	Reed et al. in 2013 [14](2),
AMACR (α-Methylacyl-CoA Racemase)	Parekh et al. in 2007 [9](2),
	Sardana et al. in 2008 [12](6),
	Reed et al. in 2010 [11](4),
	Martin et al. in 2012 [6](2),
	Prensner et al. in 2012 [10] (4),
	Velonas et al. in 2013 [14](7)
TMPRSS2: ERG/ETV1 (Gene fusion)	Parekh et al. in 2007 [9](1),
	Sardana et al. in 2008 [12](4),
	Reed et al. in 2010 [11](14),
	Martin et al. in 2012 [6](10),
	Prensner et al. in 2012 [10] (8)

Although some of the biomarkers were listed as proteins by the reviewers, we have listed the corresponding gene symbol so that each biomarker may be searched for within our web portal described in Sect. 3.3

To the best of our knowledge, there are no reviews to identify emerging biomarkers based on how many papers are being published on each biomarker. All of the reviews that we found identified emerging biomarkers by qualitative review of primary studies, where the number of references cited was no more than 14. The reviews are manually done, and none of the reviews utilized text mining to the best of our knowledge to examine the popularity of the biomarkers. Also, none of the reviews offered an analysis of the popularity of the biomarkers with time. The only biomarker that has been described in reference to its popularity with time is the famous PSA protein/KLK3 gene. For example, Prensner et al. [10] described in their review that PSA was first studied in the 1970s and in 1986 the US Food and Drug Administration (FDA) approved PSA as a test for prostate cancer in men over 50. In our study, we look at the timeline of biomarkers to see how their popularity has changed through time, and whether the biomarkers labelled as emerging are actually increasing in popularity or if their popularity is declining.

In our study, we also look at how the popularity of specific gene families has changed through time. For example, reviews done by Borgoño and Diamandis [1], Diamandis and Yousef [2], and Paliouras et al. [8] have stated that the Kallikrein gene family has gained importance in the prostate cancer field. Researchers have started to examine the role of other Kallikreins in prostate cancer in addition to KLK3. For example, KLK2 was mentioned as an emerging biomarker by four reviewers (Table 1). Similarly, other reviewers such as Vindrieux et al. [15] have identified the family of chemokines to have an important role in prostate cancer. Another gene family that has been identified as emerging in the field of prostate cancer is the S100 Protein Family, which is a family of calcium-binding proteins [14]. In our study, we seek to answer the question of whether specific gene families are emerging in popularity, or decreasing in popularity. Using our approach, we may also find whether a gene family is famous in prostate cancer for a particular gene, or for a range of genes.

3 Approach

3.1 Data Set

The number of abstract IDs collected was 134,904 on March 31st, 2016 using the keyword "prostate cancer." The abstract IDs were collected from the PubMed database using the ESearch service.[1] Using the Becas API[2] [7], we collected the titles, publication dates, and journal names of 134,707 of the articles, but for 197 of the articles the information was not available in PubMed. We used Becas for biomedical entity recognition, but we also could have used other publicly available

---

[1]http://eutils.ncbi.nlm.nih.gov/entrez/eutils/esearch.fcgi.
[2]http://bioinformatics.ua.pt/becas/api/pubmed/annotate/.

biomedical entity annotators such as PubTator,[3] TagTog,[4] and BRAT.[5] There is also NCBO BioPortal, but it only recognizes words such as "protein" or "prostate cancer," but not specific types of proteins or genes. In future work, we will further explore which biomedical annotator is best to use for the domain of prostate cancer.

We also collected the entities recognized by Becas for 132,384 of the 134,707 articles since entities for 2323 of the articles were either not existent or could not be retrieved. Therefore, we further analyzed 132,384 articles about prostate cancer. Each entity recognized by Becas was annotated with IDs corresponding to various databases. One type of entity that was collected represented genes and proteins which have the semantic group name "PRGE" as defined by Becas. All of the PRGE entities were tagged with a UniProt ID, so therefore we used the UniProt ID to retrieve the corresponding gene families and chromosomal locations through Hugo.[6]

The entities recognized by Becas were classified under different semantic groups, and Table 2 shows the breakdown of unique entities we collected under each semantic group. The top 10 entities recognized in the anatomy semantic group consisted of variations of the following words: prostate, cells, prostatic, PCa, serum, and cancer cells. The top 10 entities recognized in the disorders semantic group consisted of variations of prostate cancer, cancer, PSA, PCa, tumor, and recurrence. As a note of caution, PSA was recognized as a disorder and matched to Salivary Gland Pleomorphic Adenoma in the Unified Medical Language System (UMLS) database, although more than likely the acronym should have been matched to PSA.

Table 3 shows the top 10 most frequently mentioned genes in the abstracts. We manually removed the topmost frequently mentioned gene, NPEPPS, because it was

Table 2 The number of unique entities recognized under each semantic group

Semantic group	Number of entities
Anatomy	8503
Disorders	7204
Genes and proteins	5986
Chemicals	4599
Biological processes	3354
Pathways	1703
Cellular components	847
Species	628
Molecular functions	438
Enzymes	203
miRNA	87

---

[3] http://www.ncbi.nlm.nih.gov/CBBresearch/Lu/Demo/PubTator/index.cgi.

[4] TagTog: https://www.tagtog.net/.

[5] http://brat.nlplab.org/index.html.

[6] http://rest.genenames.org/fetch/.

Table 3 The top 10 most frequently mentioned genes in prostate cancer

Symbol	Name	# abstracts	Family
KLK3	Kallikrein related peptidase 3	13,736	Kallikreins
AR	Androgen receptor	4719	Nuclear hormone receptors
MSMP	Microseminoprotein, prostate associated	2029	N/A
KLK2	Kallikrein related peptidase 2	1879	Kallikreins
SLC20A2	Solute carrier family 20 member 2 6	1635	Solute carriers
CDKN2A	Cyclin-dependent kinase inhibitor 2A	1633	N/A
TGM4	Transglutaminase 4	1600	Transglutaminases
GNRH1	Gonadotropin releasing hormone 1	1595	Endogenous ligands
ACAD9	Acyl-CoA dehydrogenase family member 9	1585	Acyl-CoA dehydrogenase family mitochondrial respiratory chain complex assembly factors
CEP290	Centrosomal protein 290	1361	Bardet–Biedl syndrome associated

Table 4 The top 10 gene families and the corresponding number of abstracts that mentioned them

Gene family name	# abstracts
Kallikreins	15,275
Endogenous ligands	8720
CD molecules	8173
Nuclear hormone receptors	6371
Protein phosphatase 1 regulatory subunits	3942
Solute carriers	3008
SH2 domain containing	2878
Pleckstrin homology domain containing	2245
Immunoglobulin-like domain containing	2085
Basic helix-loop-helix proteins	2070

recognized incorrectly by Becas. NPEPPS is the gene symbol for aminopeptidase puromycin sensitive which coincidentally has the same acronym as PSA. The remaining top 10 genes are assumed to be correctly recognized, as they are well-known to have a relation to prostate cancer. In Table 3, we also show the gene families for each of the top genes. Table 4 shows the top 10 most frequently mentioned gene families, which do not all correspond with the gene families from Table 3. For example, CD molecules are the third most frequently mentioned gene family, yet none of the top 10 most frequently mentioned genes are part of the CD molecules gene family. One reason that CD molecules are mentioned frequently is that it contains many different genes within the family that contribute to its count. In fact, CD molecules contain the most number of genes out of the gene families that we collected (Table 5). Therefore, in our analysis we will also consider the gene families that are emerging or demerging, and the genes that make up the families.

Table 5 The top 10 gene
families with the most genes
mentioned

Gene family name	Genes mentioned
CD molecules	279
Zinc fingers C2H2-type	214
Solute carriers	195
Endogenous ligands	190
Immunoglobulin-like domain containing	130
EF-hand domain containing	116
Ring finger proteins	95
Protein phosphatase 1 regulatory subunits	88
WD repeat domain containing	87
Pleckstrin homology domain containing	87

3.2 Analysis

Our goal was to differentiate between genes/gene families that are emerging from those that are demerging. First, we calculated the MLRA values in Sect. 3.2.1 and then we weighted the years in Sect. 3.2.2. Finally, we calculated the emerging/demerging scores in Sect. 3.2.3 and concluded the memberships in Sect. 3.2.4.

3.2.1 Calculating Moving Linear Regression Angle Values

MLRA stands for Moving Linear Regression Angle [17]. This technique creates a new time series whose values are more appropriate for detecting local extrema that will help us differentiate between emerging and demerging periods.

The MLRA algorithm requires two parameters. The first parameter is the sliding window size. Each calculated MLRA value represents the trend in the last window size years. In our system, we allow the users to decide on the window size based on their definition of how many years should be averaged to find if the trend is emerging or demerging. The second parameter is the sensitivity of the algorithm. The lower the sensitivity, the less variations in the trend it can detect. In our system, we hardcoded the value to make the algorithm very sensitive to any change in the values.

3.2.2 Weighting the Years

Earlier years are not as important as later years. For example, year 2015 is more important than year 2009. Therefore, we decided to weight the years based on the window size selected in calculating the MLRA values. We grouped the years in

Group	3		2				1			
Expected Weight	0.25		0.5				1			
Year	2006	2007	2008	2009	2010	2011	2012	2013	2014	2015
Max Year - Year	9	8	7	6	5	4	3	2	1	0
n = (Max Year − Year) / (Window Size)	2.25	2	1.75	1.5	1.25	1	0.75	0.5	0.25	0
n =⌊n⌋	2		1				0			
2^{-n}	$2^{-2} = \frac{1}{2^2} = 0.25$		$2^{-1} = \frac{1}{2^1} = 0.5$				$2^{-0} = \frac{1}{2^0} = 1$			

Fig. 1 Step by step for calculating the weights of the years with a widow size = 4

descending order by the window's size and we used the division sequence of 2^{-n} to weight the groups since this choice produced groups of reasonable sizes as shown in Fig. 1. Groups that contain later years have smaller n value than those that contain earlier years, thus they have bigger weight. The weighting equation is presented in Eq. (1).

$$\text{Weight} = 2^{-\left\lfloor \frac{\text{MaxYear}-y}{\text{WindowSize}} \right\rfloor} \tag{1}$$

To better explain the equation, we present an example. Suppose we have the following sequence: YEARS = {2006, 2007, 2008, 2009, 2010, 2011, 2012, 2013, 2014, 2015}. Figure 1 presents the step by step for calculating the weights of the years.

3.2.3 Calculating the Emerging/Demerging Scores

After calculating the MLRA values and the weights of the years, we are ready to calculate the emerging and demerging scores. The emerging and demerging scores are presented in Eqs. (2) and (3), respectively. The emerging score will only take into consideration the years where the MLRA values are greater than zero. On the other side, the demerging score will only take into consideration the years where the MLRA values are less than or equal to zero.

$$\text{Emerging Score} = \sum_{y=\text{MinYear}}^{\text{MaxYear}} |\text{MLRA}_y| * \begin{cases} 0 & \text{MLRA}_y \leq 0 \\ 2^{-\left\lfloor \frac{\text{MaxYear}-y}{\text{WindowSize}} \right\rfloor} & \text{MLRA}_y > 0 \end{cases} \tag{2}$$

$$\text{Demerging Score} = \sum_{y=\text{MinYear}}^{\text{MaxYear}} |\text{MLRA}_y| * \begin{cases} 2^{-\left\lfloor \frac{\text{MaxYear}-y}{\text{WindowSize}} \right\rfloor} & \text{MLRA}_y \leq 0 \\ 0 & \text{MLRA}_y > 0 \end{cases} \tag{3}$$

3.2.4 Calculating the Memberships

The membership values for the demerging and emerging status can be calculated by normalizing the scores so that they sum up to 1. The membership equations are presented in Eqs. (4) and (5). The gene/gene family gets assigned with the status of the highest membership.

$$\text{EmergingMembership} = \frac{\text{EmergingScore}}{\text{EmergingScore} + \text{DemergingScore}} \tag{4}$$

$$\text{DemergingMembership} = \frac{\text{DemergingScore}}{\text{EmergingScore} + \text{DemergingScore}} \tag{5}$$

3.3 Web Portal

The web portal for exploration of the emerging/demerging genes or gene families can be found online.[7] The web portal currently displays information that was collected on March 31st, 2016. The front page of the web portal has two lists: the most frequently mentioned prostate cancer genes on the left-hand side, and on the right-hand side display the most frequently mentioned prostate cancer gene families in Fig. 2. Both lists also display the number of abstracts that mentioned each gene or gene family. By clicking on the icon that resembles a bar chart, the user can further explore the gene/gene family mentions per year.

As an example, the summary for the gene mentions per year for KLK3 is shown in Fig. 3. In the top right-hand corner, a summary related to the rank of the gene is shown. Since KLK3 is the most frequently mentioned gene in prostate cancer (since it codes for the widely used PSA biomarker), the KLK3 gene has a rank of 1. The summary also shows the number of papers that mention KLK3, which in this case is 13,464. Under the section titled Data in Fig. 3, the breakdown of KLK3 mentions is shown by year in the table. The breakdown of the number of KLK3 mentions per year is also shown in the visualization on the right-hand side by the orange bar chart, where the number of abstracts is the y-axis on the right-hand side. The bar chart was created using amCharts.[8] In the Data table, we display the percent rate change between consecutive years as well as the direction. For example, the number of KLK3 mentions decreased by 1.94% in 2015 as compared to the year 2014 (Fig. 3).

However, since the percent change in gene mention per year is not a good description of the gene's overall trend, the series of MLRA values are also shown as a blue line graph. The MLRA value on y-axis is located on the left-hand side of the

---

[7]http://genetimeline.alhajj.ca/.

[8]https://www.amcharts.com/javascript-charts/.

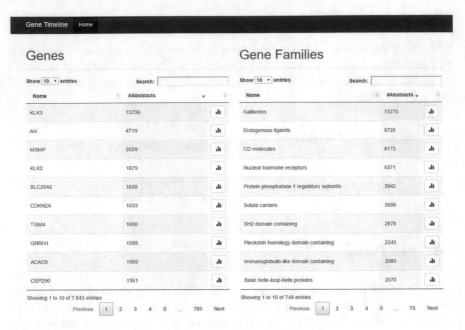

Fig. 2 The front page of the web portal where searches are initiated

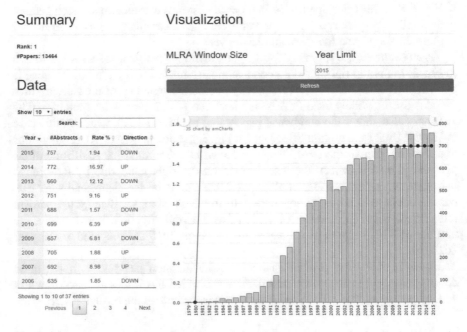

Fig. 3 The summary shown on the web portal for each gene or gene family. In this case, the summary of KLK3 is shown

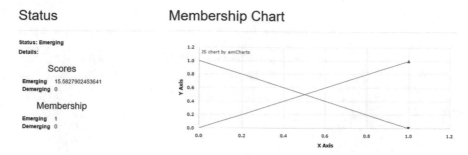

Fig. 4 The status of KLK3 is shown on the web portal

plot area. The MLRA window size (in years) can be adjusted by the user as well as the year limit (inclusive). The window size allows the user to adjust the number of previous years that are used to calculate each moving average point. The year limit allows the user to see the trend of a gene or gene family up to a specific year, in case he or she wants to compare to another study that has been written at that time.

Based on the MLRA values for KLK3, we also compute whether KLK3 is emerging or demerging, as described in Sect. 3.2. The score of KLK3 for the Emerging status or the Demerging status is calculated (Fig. 4). In this case, KLK3 has a score of approximately 15.58 for the Emerging status, and a score of 0 for the Demerging status. The membership chart is shown on the right-hand side for the Emerging status (green) and Demerging status (orange) (Fig. 4). Based on the scores and the membership chart, the membership value of KLK3 to the Emerging status is 1 and the membership value of KLK3 to the Demerging status is 0. The membership values of KLK3 are denoted on the membership chart using triangle markers. Overall, the KLK3 gene is classified as Emerging with a confidence of 100%, because the yearly publications on KLK3 are increasing each year.

We can also explore the gene family of KLK3, as shown in the screenshot of Fig. 5. At the top of the Data table in Fig. 5, the total number of papers written on Kallikreins is shown (14,937), which is a sum of all the papers listed below the first row. KLK3 comprises the largest fraction of the papers written about Kallikreins, as 13,736 of the 14,937 papers are about KLK3. Each of the Kallikrein genes are represented as a bar color within the visualization on the right-hand side, where the y-axis represents the number of papers written that mention the gene. The total amount of Kallikrein papers are represented by the gray line. On hover above each year, sorted bubbles show the breakdown of the Kallikrein papers by gene.

Finally, we can explore the co-occurrence network between the genes within a gene family, such as the genes within the Kallikrein family (Fig. 6). In Fig. 6, each edge represents co-occurrence between the two genes within an abstract, and each labelled circle represents a gene. The circumference of each circle represents the number of abstracts that have mentioned the gene. In Fig. 6, KLK3 is the most popularly mentioned gene that has often been mentioned in conjunction with the other Kallikreins. However, we see that KLK12 has never been mentioned with KLK3 together in an abstract.

Data Visualization

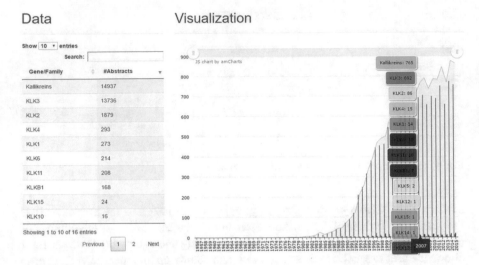

Fig. 5 Investigation of the gene family of KLK3 (Kallikreins) is shown

4 Results and Discussion

We used our approach and web portal to evaluate emerging biomarkers identified by related work. The first biomarker that we examined was KLK2, which is a serine protease from the same family as KLK3 (PSA). There were 1809 papers that had mentioned KLK2 in the abstract, which brought KLK2 to rank 4 among the other genes. The first paper to mention KLK2 was published in 1987. Using a smaller window size of 2 years, KLK2 had a membership of 0.82 to the Emerging status and a membership of 0.18 to the Demerging status. Using an MLRA window size of 5 years, KLK2 had a membership of 0.97 to the Emerging status and a membership of 0.03 to the Demerging status. Using an MLRA window size of 10 years, KLK2 has a membership of 1 to the Emerging status and a membership of 0 to the Demerging status. As a whole, KLK2 was classified as Emerging, which agreed with the related work that had classified KLK2 as emerging.

The second biomarker that we examined was transforming growth factor beta-1 (TGF-β 1), which is a growth factor involved in cell differentiation and proliferation. On our portal, TGF-β 1 was mentioned in 602 abstracts, which brought TGF-β 1 to a rank of 59. The first paper to mention TGF-β 1 was published in 1987. Using an MLRA window of 2 years, TGF-β 1 can barely be classified as emerging, as it contains a membership of 0.52 to the Emerging class, and a membership of 0.48 to the Demerging class. An increase to the window size (5 years) pushed TGF-β 1 to be classified as Demerging, as it contains a membership of 0.61 to the Demerging class and a membership of 0.39 to the Emerging class (Fig. 7). Our system showed that TGF-β 1 is losing popularity with researchers, which can be visually seen in Fig. 7 from 2011 and onwards.

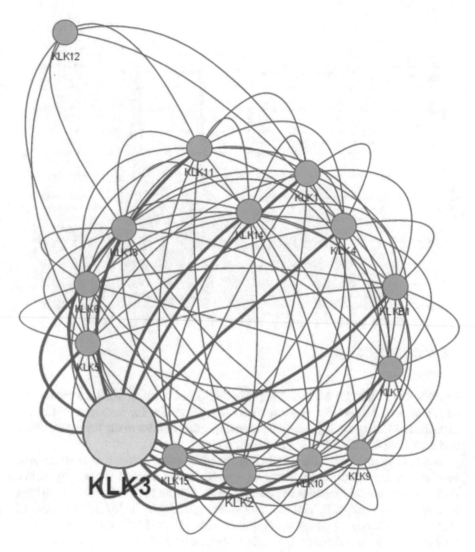

Fig. 6 The co-occurrence network between the genes of the Kallikreins family

The next biomarker we examined was Histone-lysine N-methyltransferase (EZH2), which is a polycomb group protein that is involved in the regulation of gene expression. There were only 143 papers that mentioned EZH2, where the earliest paper was published in 2001. The rank of EZH2 among the other genes is 402. With an MLRA window size of 5 years, EZH2 is indeed classified as Emerging with a membership value of 1.

We then examined interleukin-6 (IL6), which is a cytokine involved in the differentiation of B-cells. IL6 was more popular than EZH2, with 721 papers that mentioned IL6 in the abstract. The first abstract that mentioned IL6 was published

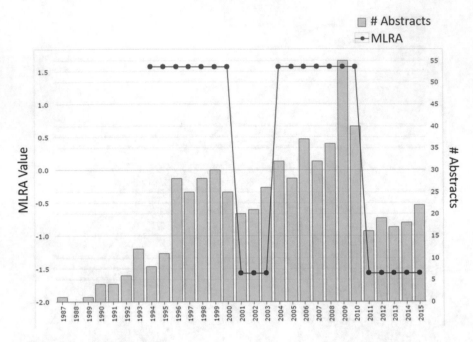

Fig. 7 The trend of TGF-β 1

in 1989, and its rank is now 46 among the other genes. With an MLRA window size of 5, which we had decided was a reasonable window size for most genes, IL6 was classified as Emerging. IL6 had a membership value of 0.93 to the Emerging class, and a membership of 0.07 to the Demerging class.

Next, we examined α-methylacyl-CoA racemase (AMACR) which is an enzyme involved in oxidative metabolism and synthesis of branched-chain fatty acids. AMACR was a newly mentioned gene, since there were only 222 abstracts written that mentioned AMACR, and the oldest was published in 2002. The rank of AMACR is 222 among the genes. However, although AMACR was mentioned by six reviewers as an emerging biomarker, the number of papers on AMACR has severely gone down from 28 in 2007 to 7 in 2015 (Fig. 8). Using an MLRA window size of 5 years, the membership value of AMACR to the Demerging status was 0.73, and its membership value to the Emerging status was 0.27 (Fig. 8). Although mentioned as an emerging biomarker by six reviewers, our web portal showed that there are fewer and fewer papers being published on AMACR.

Finally, we examined the TMPRSS2:ERG/ETV1 gene fusion signature which has been associated with prostate cancer patients. Our portal currently only allows the viewer to investigate one gene at a time unless they are part of the same family, so we could not find the frequency with which TMPRSS2 had been mentioned with either ERG or ETV1. However, since both ERG and ETV1 are part of the ETS transcription factor family, we found through our portal that ERG is the most frequently mentioned gene in that family with 601 abstracts. The second most

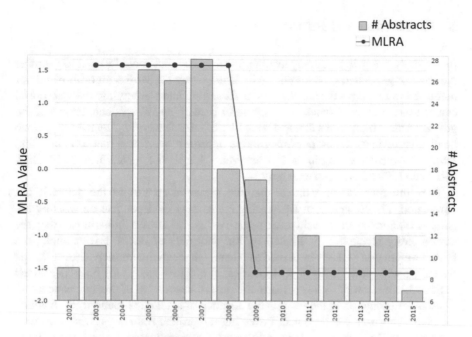

Fig. 8 The trend of AMACR

mentioned gene in the ETS transcription family was ETV3 (332 papers), then ELF3 (80 papers), and then ETV1 (75 papers). Therefore, ERG and ETV1 could have been mentioned together in at most 75 papers. We found from the co-occurrence network that ERG had been mentioned with ETV1 in some abstracts. Interestingly, ETV3 has been exclusively mentioned with ERG in abstracts.

ERG was first mentioned in 1993, but there have since been 556 papers written that mentioned ERG in the abstract. The rank of ERG is 66 among the genes. Using an MLRA window size of 5 years, ERG is indeed an emerging biomarker, with a membership value of 1 to the Emerging status. With similar parameters, ETV1 is barely emerging, with a membership value of 0.60 to the Emerging status. However, ETV1 is much newer than ERG, as it was first mentioned in 2005 and since then 73 papers have mentioned ETV1 in the abstracts. The rank of ETV1 is 611. The second gene of the gene fusion is TMPRSS2, which has a rank of 57, has 628 abstracts, and was first mentioned in 1983. It is reasonable that TMPRSS2 has more abstracts than either ERG or ETV1, as TMPRSS2 may be mentioned exclusively with either ERG or ETV1. Using an MLRA window of 5 years, TMPRSS2 has a membership value of 1 to the Emerging status. TMPRSS2 is also the most popular serine protease in prostate cancer.

5   Conclusion and Future Work

In our study, we built a publicly available web portal that could be used to examine the trend of genes and gene families with time. Using the MLRA algorithm and our membership functions for the statuses of Emerging and Demerging, the web portal can be used to classify genes as emerging or demerging. We evaluated some of the prostate cancer biomarkers that had been identified as "emerging" by manually done qualitative reviews. Using our quantitative approach, we found that TGF-β 1 and AMACR were demerging in the last few years, but KLK2, EZH2, IL6, TMPRSS2, ERG, and ETV1 have been emerging.

One limitation of the work is that the semantic context of the genes is not considered. The Emerging or Demerging status may be dependent on whether the gene is mentioned in a positive or negative light. Another limitation is that the named entity recognition is limited by the shortfalls of Becas. For example, the PSA acronym was wrongly identified as puromycin-sensitive aminopeptidase. In the future, we will evaluate different biomedical annotators such as the ones mentioned in Sect. 3.1 to see which ones perform the best in the context of prostate cancer.

Another limitation of the work is the absence of available ground truth to which the above approach can be tested. Providing the ground truth and testing it against the results of the current mode is an interesting research topic in itself and it could be combined with a second time parameter to assess the importance of this approach over time since new trends in who is publishing and what is published may change. For example, the new trend in physics is to publish rapid research, not refereed, in arXiv [4].

We will continue to make improvements to the web portal so that researchers may use it to see the standing of the genes or gene families that they are studying. One improvement that we will make is to allow users to compare multiple genes, even if those genes are not in the same family. Comparing multiple genes may be useful to find the standing of gene signatures or gene fusions.

References

1. Borgoño, C.A., Diamandis, E.P.: The emerging roles of human tissue kallikreins in cancer. Nat. Rev. Cancer **4**(11), 876–890 (2004)
2. Diamandis, E.P., Yousef, G.M.: Human tissue kallikreins: a family of new cancer biomarkers. Clin. Chem. **48**(8), 1198–1205 (2002)
3. Draisma, G., Boer, R., Otto, S.J., van der Cruijsen, I.W., Damhuis, R.A., Schröder, F.H., de Koning, H.J.: Lead times and overdetection due to prostate-specific antigen screening: estimates from the European randomized study of screening for prostate cancer. J. Natl. Cancer Inst. **95**(12), 868–878 (2003)
4. Economist: Time's up. Economist **422**(9033), 69–71 (2017)
5. Etzioni, R., Penson, D.F., Legler, J.M., di Tommaso, D., Boer, R., Gann, P.H., Feuer, E.J.: Overdiagnosis due to prostate-specific antigen screening: lessons from us prostate cancer incidence trends. J. Natl. Cancer Inst. **94**(13), 981–990 (2002)

6. Martin, S.K., Vaughan, T.B., Atkinson, T., Zhu, H., Kyprianou, N.: Emerging biomarkers of prostate cancer (review). Oncol. Rep. **28**(2), 409–417 (2012)
7. Nunes, T., Campos, D., Matos, S., Oliveira, J.L.: Becas: biomedical concept recognition services and visualization. Bioinformatics **29**(15):1915–1916 (2013)
8. Paliouras, M., Borgono, C., Diamandis, E.P.: Human tissue kallikreins: the cancer biomarker family. Cancer Lett. **249**(1), 61–79 (2007)
9. Parekh, D.J., Ankerst, D.P., Troyer, D., Srivastava, S., Thompson, I.M.: Biomarkers for prostate cancer detection. J. Urol. **178**(6), 2252–2259 (2007)
10. Prensner, J.R., Rubin, M.A., Wei, J.T., Chinnaiyan, A.M.: Beyond PSA: the next generation of prostate cancer biomarkers. Sci. Transl. Med. **4**(127), 127rv3 (2012)
11. Reed, A.B., Parekh, D.J.: Biomarkers for prostate cancer detection. Expert Rev. Anticancer Ther. **10**(1), 103–114 (2010)
12. Sardana, G., Dowell, B., Diamandis, E.P.: Emerging biomarkers for the diagnosis and prognosis of prostate cancer. Clin. Chem. **54**(12), 1951–1960 (2008)
13. Stamey, T.A., Yang, N., Hay, A.R., McNeal, J.E., Freiha, F.S., Redwine, E.: Prostate-specific antigen as a serum marker for adenocarcinoma of the prostate. N. Engl. J. Med. **317**(15), 909–916 (1987)
14. Velonas, V.M., Woo, H.H., Remedios, C.G.d., Assinder, S.J.: Current status of biomarkers for prostate cancer. Int. J. Mol. Sci. **14**(6), 11034–11060 (2013)
15. Vindrieux, D., Escobar, P., Lazennec, G.: Emerging roles of chemokines in prostate cancer. Endocr. Relat. Cancer **16**(3), 663–673 (2009)
16. Wang, M., Valenzuela, L., Murphy, G., Chu, T.: Purification of a human prostate specific antigen. Investig. Urol. **17**(2), 159–163 (1979)
17. Xylogiannopoulos, K., Karampelas, P., Alhajj, R.: Discretization method for the detection of local extrema and trends in non-discrete time series. In: ICEIS 2015 - 17th International Conference on Enterprise Information Systems (2015)

Exploring the Role of Intrinsic Nodal Activation on the Spread of Influence in Complex Networks

Arun V. Sathanur, Mahantesh Halappanavar, Yi Shi, and Yalin Sagduyu

Abstract In many complex networked systems, such as online social networks, activity originates at certain nodes and subsequently spreads on the network through influence. In this work, we consider the problem of modeling the spread of influence and the identification of influential entities in a complex network when nodal activation can happen via two different mechanisms. The first mechanism of activation stems from factors that are intrinsic to the node. The second mechanism comes from the influence of connected neighbors. After introducing the model, we provide an algorithm to mine for the influential nodes in such a scenario by modifying the well-known influence maximization algorithm. We sketch a proof of the submodularity of the influence function under the new formulation and demonstrate the same on larger graphs. Based on the model, we explain how influential content creators can drive engagement on social media platforms. Using additional experiments on a Twitter dataset, we then show how the formulation can be applied to real-world social media datasets. Finally, we derive a centrality metric that takes into account both the mechanisms of activation and provides for an accurate, computationally efficient, alternate approach to the problem of identifying influencers under intrinsic activation.

Keywords Complex networks · Influence maximization · Social influence · Intrinsic activation · Independent cascade · Monte carlo · Centrality · Spectral methods

A. V. Sathanur (✉)
Pacific Northwest National Laboratory, Seattle, WA, USA
e-mail: arun.sathanur@pnnl.gov

M. Halappanavar
Pacific Northwest National Laboratory, Richland, WA, USA
e-mail: hala@pnnl.gov

Y. Shi · Y. Sagduyu
Intelligent Automation, Inc., Rockville, MD, USA
e-mail: yshi@i-a-i.com; ysagduyu@i-a-i.com

© Springer International Publishing AG, part of Springer Nature 2018   123
M. Kaya et al. (eds.), *Social Network Based Big Data Analysis and Applications*,
Lecture Notes in Social Networks, https://doi.org/10.1007/978-3-319-78196-9_6

1 Introduction and Related Work

The advent and rapid adoption of social media platforms allows people to self-organize into complex social networks with rich dynamics. Users can disseminate their views, opinions, and other content while simultaneously consuming and reacting to the content created by friends, people, and organizations they follow. The success of such platforms depends on the myriad of content creators, the quality of their content, and the activities their audiences generate because of the various types of engagement possible with the posted content. These actions can be attributed to influence. The dynamics of influence and resulting diffusion of information in complex networks has been the subject of intense scrutiny for researchers and practitioners in many fields with particular attention to the identification of central or influential nodes on the network. One rigorous approach to finding influential users with motivations originating in viral marketing is the approach based on *influence maximization*.

We can define the influence maximization problem as follows: Consider a directed graph $G = (V, E)$ that abstracts a complex network, where V is the set of nodes $V = \{v_1, v_2, v_3 \ldots\}$ and E is the set of directed edges $\{(v_u, v_w) | v_u, v_w \in V\}$. The directed edge (v_u, v_w) implies that v_u can influence v_w and not the other way round. However, it is possible that both (v_u, v_w) and (v_w, v_u) are valid edges. We denote by $|V|$ the total number of nodes and by $|E|$ the total number of edges in the graph G. Further, the nodes are labeled as either *Passive* or *Active*, denoting the state of the vertex. A necessary but not sufficient condition for an active vertex v_u to activate a passive vertex v_w is that $(v_u, v_w) \in E$. Other conditions come from the nature of the diffusion model. Given that it is possible to initially activate k nodes, the influence maximization problem aims to find the particular set of k *seed nodes*, called the *seed set S*. When the nodes in the set S are activated, the spread of influence results in maximal activations on the network among all possible such sets of k nodes. Note that in the subsequent discussions, we use the terms *reachability*, *number of activations on the network*, and *influence spread* synonymously to denote the total number of active nodes on the network after running the diffusion models, starting from the initial set of active nodes, until no more activations are possible.

Starting with the landmark paper by Kempe et al. [10], several works have explored newer diffusion models and variations to the ones studied in the work by Kempe et al., namely, the independent cascade (IC) model and the linear threshold (LT) model. These models explicitly address the various sociological aspects of influence. Li et al. in [15] consider influence dynamics and influence maximization under a general voter model with positive and negative edges. In a follow-up work, Kempe et al. [11] discuss a diverse set of models including the so-called decreasing cascade model where attempts by multiple neighbors to activate a node result in decreasing probability of activation, as the size of the set of neighbors trying to activate the node increases. The authors in Ref. [23] propose a general diffusion model that takes into account different granularities of influence, namely, pair-wise, local neighborhood, etc. The authors in [4], consider influence maximization under the scenario where negative opinions may emerge and propagate. In [8], the authors consider the problem of identifying the individuals whose strong positive opinion about a product will maximize the overall positive opinion about the product. In the process, the authors leverage the social influence model proposed by Friedkin and Johnsen [7]. For a comprehensive survey on the various models of influence, we refer the reader to the paper by Zhang et al. [26].

Next, we consider the models that address two different types of activation: intrinsic and influenced. The interplay of these two mechanisms is exemplified by three different scenarios outlined below.

- Users posting content on social media due to their own initiative constitutes intrinsic activation. Actions such as sharing, retweeting, commenting constitute influenced activation.
- Posting behavior that is external to a given network can be considered to be intrinsic activation. This would include watching a video on a website from a shared e-mail link and then sharing it on Twitter. From the perspective of just the Twitter network, it appears that such users are intrinsically activated.
- In a traditional social network, such as a physical community, that is not an OSN (online social network), intrinsically activated users would be those who take the initiative to start an activity, for example, a campaign for social good. The same can then spread through word of mouth, flyers, etc.

Myers, Zhu, and Leskovec investigate the diffusion of information, with origins external to that of a social network, through the internal social influence mechanism [17]. In a recent work [6], the authors recognize that the events on social media can be categorized as exogenous and endogenous and model the overall diffusion through a multivariate Hawke's process to address activity shaping in social networks. In another recent work, the authors in [18] propose a novel diffusion model based on factor graphs and graphical models where the node potentials can correspond to the notion of intrinsic activation in our case. However, the focus of their work is on the diffusion model itself, not on the aspects of intrinsic activation. While being similar in spirit to these works, our work is geared toward modeling the spread of influence and mining influential nodes in scenarios with intrinsic and influenced activation-aspects that have not been studied in existing literature.

We make the following contributions in this work:

1. Our approach results in a probabilistic model for two different types of nodal activations, namely, intrinsic and influenced mechanisms found in real-world networked systems, such as OSNs.
2. We examine these mechanisms in the context of influencer mining from *two different perspectives*: the well-known combinatorial influence maximization perspective and a generalized centrality perspective.
3. We define a modified influence spread function, sketch a proof of its submodularity, and provide a modified version of the influence maximization algorithm to maximize the new influence spread function.

4. We examine the nature of content creators and consumers on a social network in light of the two activation mechanisms.
5. Carefully chosen experiments on synthetic and real-world graphs are used to illustrate various aspects of the model and compare it to the independent cascade model.
6. We derive a new centrality metric from the activation model and show that this metric can accurately identify influential users in a computationally efficient manner.

The initial aspects of this work was published in [21]. The present version is a significant extension of the above work where we have extensively examined the content creation and content spreading mechanisms, formally sketched a proof of the submodularity of the modified influence function, added an extensive set of experiments on a real-world Twitter graph, and improved the overall narrative by means of several smaller additions.

2 Modified Influence Maximization Approach

2.1 Formulation

Considering that nodal activation can originate from two different mechanisms, *Intrinsic* and *Influenced*, allows us to effectively model the so-called *self-evolving* systems (e.g., OSNs) that are comprised of content creators (higher probability of getting activated intrinsically) and content spreaders (activated via influence and spreading the information). Recognizing that most of the users are in some sense both activity creators and content spreaders (typically also the content consumers) at the same time, we introduce a real-valued parameter $\alpha \in [0, 1]$ that models the probability of self-activation. The total probability for activation of a given node (user) i is composed of the probability of activation from the two different mechanisms. The parameter $\alpha(i)$ denotes the probability of intrinsic activation, and $\beta(i)$ denotes the probability through influence with $\alpha(i) \geq 0$ and $\beta(i) \geq 0$. The influenced part of the probability for activation is comprised of the activation probabilities due the 1-hop neighbors of the user under consideration.

Note that there are many interaction models that are studied under influence maximization as pointed out in Sect. 1. Our model with intrinsic activation is based on one of the most widely studied models, namely, the independent cascade model, and this will be the focus of the current work. Specifically, in this work, we do not consider developing the variants of other models incorporating intrinsic activation. Similar to the IC model, the weights w_{ij} ($0 \leq w_{ij} \leq 1$) when multiplied by $\beta(i)$ denote the probability of user j activating user i, given that user j is activated by either of the above means. Figure 1 describes these mechanisms and the associated

Fig. 1 A concise representation of the self and influenced mechanisms of activation of a node i

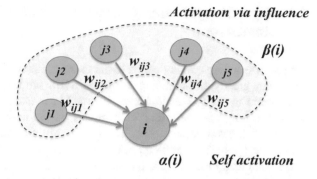

coefficients. The described probabilistic formulation has similarities to the Friedkin-Johnsen social influence model for opinion change [7] where the authors recognize that the dynamics of opinion change are governed by two mechanisms-the intrinsic opinion and the influenced opinion.

We also adopt the weighted-cascade version of the IC model by normalizing the edge probabilities [10], so that the expected number of nodes influencing a given node is 1. Henceforth, in this work, when we refer to the IC model, we imply the weighted cascade version of the IC model. However, this is not a limitation of the model since our model can also be used in the pure IC model setting. Thus, if W denotes the sparse weight matrix that characterizes the IC edge probabilities, we require that W be row stochastic. That is, $\sum_{j,(j,i)\in E} w_{ij} = 1$. Further by assuming that the nodes are not *lazy* and are activated by either of the two mechanisms that we outline, we set $\beta(i) = (1 - \alpha(i))$. This will render the overall IC probability between nodes j and i to be $(1 - \alpha(i))w_{ij}$. While the changed probabilities denote a departure from the weighted-cascade model, when the effect of intrinsic activation is added back, the expected number of nodes activating a given node is still 1.

Note that all the model parameters discussed can be efficiently determined either by a maximum-likelihood-based approach (as in this work) or by alternative methods such as the expectation-maximization (EM) approach followed in reference [19]. For example, the proportion of tweets by a user i that are intrinsic in nature can quantify $\alpha(i)$, while a particular weight w_{ij} can be determined by the proportion of user i's retweets (or influenced activity) having their origin in the activity of user j that user i follows. While these *local influence models* can be determined in alternate ways, our goal is to find the overall influencers once these model parameters are estimated.

Our formulation addresses the problem of identifying influential nodes on a network without explicit seeding. The original influence maximization approach with roots in viral marketing explicitly activates the seed nodes while in our formulation, the system is self-evolving in that nodes get activated intrinsically with a probability (content creation) and subsequently these activations spread (content consumption and spreading) on the network. This is the focus of the next section which describes the modifications to the original influence maximization algorithm necessary to identify the influencers under intrinsic activation.

2.2 Algorithm for Mining Influential Nodes Under Intrinsic Activation

We propose a simple modification to the classic influence maximization framework using the greedy hill-climbing optimizer [10], working with the IC model, to incorporate the self-activation mechanism. Let us assume that we are seeking k influential nodes out of a total of N nodes on the network. Let S^p be the set of influential nodes at step $p \leq k$. The greedy hill-climbing optimizer expands the set to size $(p+1)$ by polling each of the nodes not in S^p and augmenting those nodes, one at a time to form the set $S^p \cup \{v\}$ and looking for the best marginal gain in terms of the activations. At each such step p, instead of setting each of the nodes in $S^p \cup \{v\}$ to be activated and then computing the activations according to the IC model, we probabilistically activate each node in $S^p \cup \{v\}$ with a probability given by the corresponding α values to simulate the intrinsic activation process. This modification is depicted in line 9 of Algorithm 1. Given the probabilistic nature of the algorithms, the overall activation numbers are obtained by running the diffusion model in a Monte Carlo fashion by invoking n independent trials involving randomized graphs with corresponding edge weights.

Note that this algorithm results in the computation of a modified influence spread objective function $\sigma(S)$ (same as a_{best} in the algorithm), which gives us the total number of activations on the network attributable to the multi-hop influence of

---

Algorithm 1 Selects a set of k influential nodes that cause maximal activations on a network, following the independent cascade (IC) model with self-activation (IC-Int). The inputs are a directed graph ($G = (V, E)$), set of edge probabilities ($P = \{w_{ij} : (ji) \in E\}$), vector of alpha values ($\alpha = \{\alpha_v : v \in V\}$), number of samples ($n$), and number of influential nodes to be identified (k)

1: **procedure** IC-INT(G, P, α, k, n)
2:      Generate n random numbers $r_{uv}^1 \ldots r_{uv}^n$ for each edge in E and generate a set SG containing
        n subgraphs such that in subgraph i, $w_{ij} \geq r_{uv}^i$
3:      $S \leftarrow \emptyset$ ▷ Set of influential nodes to be mined
4:      **while** $|S| < k$ **do**
5:          $v_{best} \leftarrow \emptyset, a_{best} \leftarrow 0$
6:          **for** each node v in $V \setminus S$ **do**
7:              $a \leftarrow 0$
8:              **for** each $G_i \in SG$ in parallel **do**
9:                  $\hat{S} \leftarrow$ active nodes in $S \cup \{v\}$ based on α
10:                 Compute number of nodes, $\hat{a}$, in $V \setminus \hat{S}$ that are reachable from the $\hat{S}$
11:                 $a \leftarrow a + \hat{a}$ ▷ Synchronized update
12:             **if** $a \geq a_{best}$ **then**
13:                 $v_{best} \leftarrow v$
14:                 $a_{best} \leftarrow a$
15:         **if** $v_{best} \neq \emptyset$ **then**
16:             $S \leftarrow S \cup \{v_{best}\}$
17:     **return** S
---

nodes in the set S when the corresponding nodes are activated intrinsically, in accordance with their α values. Thus, during this process, at the step denoted by line 10, the nodes in the set $(V \setminus S^P \cup \{v\})$ are not activated intrinsically, instead they are activated via influence. These aspects are discussed further in Sects. 2.3 and 2.4.

The running time of Algorithm 1 depends on the α values since they affect whether a particular node is active in the given sample or not (line 9 of the algorithm). If the node is not active, then reachability will not be computed from that node. The worst-case complexity of the algorithm is the same as that for the independent cascade model and can be derived to be $O(nk^2|V||E|)$ where n is the number of Monte Carlo samples, k is the number of influential nodes sought, $|V|$ is the number of nodes (vertices) in the graph, and $|E|$ is the number of edges in the graph. The approach to solve the problem as detailed in Algorithm 1 is based on the classic greedy algorithm to maximize monotone submodular functions. There are two ways to make the algorithm scalable. One is to accelerate the outer greedy optimization loop, and the second method is to improve the scalability of the reachability computation. There is prior work on both the areas. For example, the work presented in [14] uses lazy evaluations to speed up the greedy algorithm while reference [16] uses a stochastic version of the greedy algorithm to improve the scalability. References [5] and [3] on the other hand use techniques to speed up the reachability evaluations on the sampled subgraphs. Because our objective function is also monotone submodular (more on this in Sect. 2.4) and uses reachability computations, it can benefit from these algorithms to scale to networks with millions of nodes.

2.3 Content Creators and Engagement in Online Social Networks

With the help of the described activation model, we examine aspects of content creation, consumption, and content spreading in OSNs, and how these are tied to the success of the platform as a whole. Figure 2 shows the out-links around a source node (s) and the various receiver (follower) nodes ($r_1 \ldots r_k$) with the α and w values.

Using the described activation model, it is evident that by restricting ourselves to one hop, we can write the modified influence function $\sigma(s)$ that denotes the expected number of nodes activated by node s as follows:

$$\sigma(s) = \alpha_s \sum_{k=1}^{d_s^o} \left(1 - \alpha_{r_k}\right) w_{r_k s} \tag{1}$$

Here, d_s^o denotes the out-degree of the node s. We are only modeling the activation through influence when the source node s gets activated intrinsically.

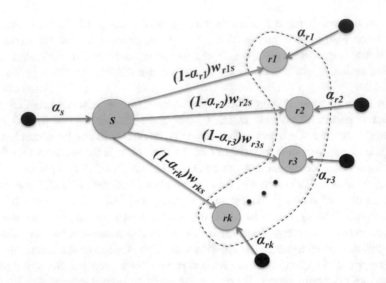

Fig. 2 The out-links from a source node (s) to the receiver nodes ($r_1 \ldots r_k$), and the associated node and edge attributes

We do not add the term $\sum_{k=1}^{d_s^o} \alpha_{r_k}$ that denotes the intrinsic activation of the nodes ($r_1 \ldots r_k$). This is because activation via influence creates engagement on the social network platform (e.g., sharing, commenting, and liking). Thus $\sigma(s)$ can be viewed as a surrogate for engagement. The set of influential nodes (users) that we wish to compute by following Algorithm 1 denotes the set of influential content creators that are able to maximize this engagement (by which we mean the influence spread under intrinsic activation of the creator nodes), and are quite valuable to the platforms.

The scenario in which all nodes have large values of α will result in nodes creating high volume of content on their own, and there is not much spreading of the content through different forms of engagement. Alternatively, all nodes having rather small values of α means that while the nodes are eager to spread the content, there is not much content created in the first place, again reducing the engagement. Therefore, we hypothesize there is an optimal assignment of the α values for a given assignment of the interaction probabilities and the network topology that can maximize the spread of influence under intrinsic activation. While we provide evidence of this with experiments on a real-world Twitter dataset in Sect. 4, solving an actual optimization problem is beyond the scope of this work.

Equation (1) provides a quick preview of the distribution of the α values that can lead to maximizing this engagement. The objective function $\sigma(s)$ favors a source node with large α_s and high out-degree, connected to receivers with low α_{r_k} values who easily engage with the intrinsic activity of the source node (higher value of the IC probability along these edges and lower value of the receiver α). In practice when users can have arbitrary α values, Algorithm 1 is able to seamlessly identify such influential content creators by simulating the two mechanisms of activation.

The same will not be possible with the independent cascade model because every node that is selected to be a part of the seed set is necessarily activated, thereby overestimating a given node's influence.

2.4 Optimality of the Influence Maximization Algorithm with Intrinsic Activation

For the classic influence maximization problem with the IC model, the greedy hill-climbing optimizer is shown to be optimal in the sense that it provides $(1 - \frac{1}{e} - \epsilon)$ approximation guarantee on the expected influence spread function. This is because the expected influence spread $\sigma(S)$ is a monotone submodular function [10, 12]. The greedy algorithm expands the seed set S by the addition of nodes with the highest marginal gain in terms of the number of activations. For the case of intrinsic nodal activation, we have nodes activated intrinsically, as well as through influence. Thus, it appears that an influence function defined by the total number of activations on the network is not submodular. However, given that we are only interested in the total number of activations caused by the spread of intrinsic activations (a_{best} in Algorithm 1), the submodularity property can be shown to remain valid.

For each node on the network $i \in V$, we can introduce an edge pointing from a newly created dummy node i_D to the actual node i with an activation probability equal to α_i. This process is illustrated in Fig. 2. Let V_D denote the set of dummy nodes. Note that there is a one-to-one correspondence between the nodes in V_D and V. Also, because every node i_D in V_D has a single outgoing edge to the corresponding node i in V, i_D cannot be activated by i. On the other hand, given a large number of samples n, the expected number of times the edge between any pair of nodes $(i_D \in V_D, i \in V)$ is activated is $n\alpha_i$ leading to us to represent the IC probability between i_D and i to be α_i. This is represented by line 9 of Algorithm 1. Thus, the original influencer mining problem can now be transformed to mining for influential nodes in the set V_D under the IC model. Given that the influence (cumulative reachability) function is submodular under the IC model [10], the influence function in the case of intrinsic activation being present, namely, a_{best} in Algorithm 1, is also submodular.

3 Synthetic Experiments

3.1 Small Organization Tree

We first consider a small directed and weighted network with 23 nodes, organized in a tree-like fashion. The graph is depicted on the left side of Fig. 3. In this experiment, we consider the tree-like network to depict a small organization with

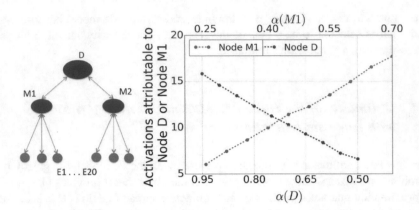

Fig. 3 The small organizational tree network (left) and the behavior of the influence functions with the various α values

a director (Node D), two managers (M1 and M2), and 20 employees (E1–E20), with 10 employees each working under the two managers. We set $\alpha^0(D) = 0.95$, signifying that the director almost exclusively acts intrinsically. We also set $\alpha^0(M1) = \alpha^0(M2) = 0.25$. All Employees have an α of 0.25 as well. As for the weights (same as the activation probabilities in the IC model), the edges ending at node D receive weights of 0.5 each (when the director chooses to be influenced, the director gets influenced equally by the two managers). As for the managers, they have a weight of 0.5 each on the edges that are incoming from D and the remaining 0.5 is split equally among the edges originating at the 10 employees each. All Employees carry a weight of 1.0 on the edges originating from the managers. We then perturb this baseline case to mimic a situation where the director starts becoming more susceptible to influence, while the manager M1 becomes inflexible. This is simulated by setting $\alpha(D) = \alpha^0(D) - \delta$ and $\alpha(M1) = \alpha^0(M1) + \delta$. We then sweep δ from 0.05 to 0.45. The results are shown in the right panel of Fig. 3, where we can see that D begins as the most influential node as expected, but then M1 becomes more influential than D at a certain value of δ and will eventually have reach over most of the employees on the network. Note that the activation numbers plotted on the y-axis are the expected numbers over a Monte Carlo analysis with $n = 3200$ samples. This simple experiment shows that the nature of influence spread on social networks is sensitive to the extent of intrinsic activation of the key nodes. Clearly, these scenarios cannot be easily captured by the IC model, where the concept of intrinsic activation with a continuous probability value ($\alpha \in [0, 1]$) does not exist.

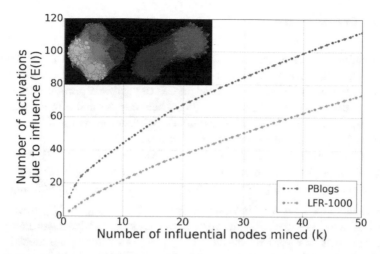

Fig. 4 Submodular nature of the influence function under self-activation. *Inset: The PBlogs (left) and LFR-1000 (right) networks visualized in Gephi [2].* x-Axis refers to the number of influential nodes mined and y-axis refers to the expected number of activations achieved due to influence. This is represented as $E(I)$

3.2 Larger Graphs and the Influence Function

Our next experiment involves two larger graphs where the topology of one is from a real-world dataset, while the other is synthesized. In both cases, the node α values are drawn from a uniform distribution $U[0, 1]$, and the w_{ij} values are also drawn from $U[0, 1]$ and then normalized as described earlier in Sect. 2.1. The graphs under consideration are described below:

- LFR-1000 graph with 1000 nodes and 11,433 edges is a synthetic network that follows the generative LFR model that mimics real-world graphs [13].
- The PBlogs graph [1] represents a real-world blogs network and has 1095 nodes and 12,597 edges.

Further details of these graphs are discussed in [9].

When we applied the modified influence maximization approach given by Algorithm 1 to the LFR-1000 and the PBlogs graphs and requested for 50 seeds, we observed (Fig. 4) that the cumulative influence spread (total number of influenced activations) showed a submodular character as evidenced by the diminishing gains in the total number of activations for each new seed added to the set.

4 Experiments on a Real-World Twitter Dataset

In this section, we consider the various aspects of intrinsic activation on an interaction graph constructed from Twitter data.

4.1   Data Collection

We first build a directed follower/friend graph from Twitter data using the public
Twitter API [24], where each user is a vertex and a directed edge (u, v) from user
u to user v means that v follows u. Our goal is to capture a portion of the Twitter
graph such that there are enough interactions between the nodes to estimate the α
and the w_{ij} parameter values with reasonable confidence as required by our model.
Algorithm 2 depicts the graph construction details.

   Algorithm 2 starts by adding a seed twitter user u_0 to the set S (line 2) and by
adding followers. In order to improve the density of the graph (as measured by
$\frac{|E|}{|V|^2}$), we pick up to k_{in} (k_{in} is set to 15 in our experiment) users with highest in-
degree values in the set S to form a new set $S' \subseteq S$. If $k_{in} > |S|$, then we will
just pick all the nodes in set S. New vertices and edges are added accordingly (lines
9–15). Note that new vertices are added if the users are being introduced for the
first time. Low out-degree nodes (based on the threshold k_{out}, which is set to 11 in
our experiments) are excluded in the graph construction (line 18). The process is
repeated until a required number of vertices have been added to G. Random seeds
are added to S when it becomes empty (line 17).

Algorithm 2 Generate a directed graph, $G = (V, E)$, from the given Twitter dataset.
Unique users in the dataset are represented as nodes, and the notion of a follower
is represented as a directed edge. If user v follows user u, then add a directed edge
(u, v). The desired number of nodes n and sampling parameters k_{in} and k_{out} are
provided as inputs to the algorithm

```
 1: procedure TWITTER-GRAPHGEN(G = (V, E), k_in, k_out, n)
 2:     S = {u_0}                                                        ▷ Seed User
 3:     V(G) ← {u_0}
 4:     E(G) ← ∅
 5:     while |V| < n do                           ▷ Graph is less than the desired size
 6:         S' ⊆ S              ▷ Select k_in users from set S such as based on top in-degree
 7:         S ← S \ S'
 8:         F ← ∅
 9:         for each user u in S' do
10:             F ← F ∪ { all followers of user u}
11:             for each node v in F do
12:                 if v ∉ V(G) then
13:                     V ← V ∪ {v}
14:                     S ← S ∪ {v}
15:                     E ← (u, v)                               ▷ Add a directed edge (u, v)
16:         if S = ∅ then
17:             S = {u_rand}                      ▷ Randomly select a user from the dataset
18:         Recursively remove nodes with out-degree < k_out. Goto Line 5
19:     return G
```

Starting with the user "PurdueEngineers" as the `seed_user` u_0 to collect data, and following Algorithm 2, we obtain a graph with 1167 nodes and 10,292 edges. We then generate an interaction graph from the follower/friend graph by assigning a weight on each edge (u, v) based on the number of interactions, where interactions refer to replies, retweets, or mentions. We define weight $\gamma_{(u,v)}$ for a directed edge (u, v) as one plus the number of times user v replies to, retweets, or mentions user u. Note that we need to define positive weights and thus we define weight prior to normalization by one plus the number of interactions.

4.2 Influence Spread Results

Given the dataset with tweets and interactions in the form of retweets, replies, and mentions, we estimate the node-specific parameters α_i and the edge-specific parameters w_{ij} as below.

$$\gamma_i = \sum_j \gamma_{(j,i)},$$

$$\alpha_i = \frac{k_i}{\gamma_i + k_i},$$

$$\beta_i = 1 - \alpha_i,$$

$$w_{ij} = \frac{\gamma_{(j,i)}}{\gamma_i}.$$

Here, k_i refers to the total number of intrinsic tweets from user i and γ_i is the total number of interactions that user i participated in. Meanwhile $\gamma_{(j,i)}$ breaks this up according to the interactions with the users that user i follows.

Note that the nature of interactions between two users can be highly complex and may be dependent on a host of features. However, in this work, we are not concerned with the complexities of the interactions. We simply compute the parameters based on counts of tweets and interactions in a *maximum likelihood manner* without regard to other features, such as topics and sentiment strength.

On the same Twitter graph, we compare the activations achieved by the IC model and the present model, incorporating intrinsic activation. We retain the interaction probabilities as the same between the two models while noting that any interaction probability w_{ij} becomes $(1-\alpha_i)w_{ij}$ for our model. We then randomized the intrinsic activation parameter α for each of the users to observe if the influenced activations can match that of the IC model over a number of trials. Figure 5 illustrates the results from a Monte Carlo analysis with 50 trials. Here the influence spread curves corresponding to the Monte Carlo runs for the model with intrinsic activation (IC-Int) are all well below the influence spread curve for the IC model as expected. Furthermore, these results are in line with the observation that the authors make in [18] where they show that the IC model significantly overestimates the activations.

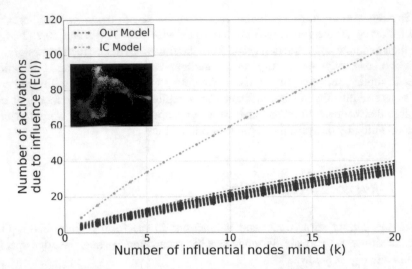

Fig. 5 Comparing the influence spread with the IC model (Green) and the Monte Carlo runs on our model with intrinsic activation (Red). *Inset: The constructed Twitter graph is visualized in Gephi*

Fig. 6 Distribution of the percentage overlap between the influencers identified by the IC model and the IC model with intrinsic activation in a Monte Carlo run

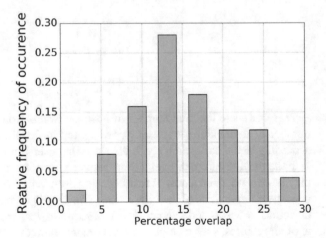

Next, for each of the Monte Carlo runs, we identify the percentage overlap between the sets of influential seeds identified by the IC model and our model with intrinsic activation. These results are shown in Fig. 6 for 50 Monte Carlo runs and for top 30 influential nodes. Note that while a small number of runs show nearly no overlap, more than 25% of the runs show 20% or more overlap. This is due to the fact that both the IC model and our model with intrinsic activation favor nodes with large out-degrees. However, the IC model with intrinsic activation also requires that such nodes have a high enough α value to be influential along the lines of the discussions in Sect. 2.3.

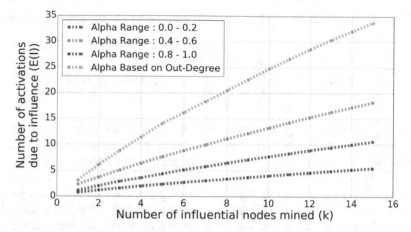

Fig. 7 The mean activation curves for three different ranges of α assignments and the fourth with α values proportional to the node out-degree

Continuing our discussions that began in Sect. 2.3 regarding maximizing the *engagement*, we consider four different cases of assigning α values to the nodes. In the first three cases, the random α values are drawn from three different intervals in a uniform manner. In the first case, all the α values are drawn according to the distribution $U[0, 0.2]$. Meanwhile in the second case, they were drawn from $U[0.4, 0.6]$ and finally in the third case, from the distribution $U[0.8, 1.0]$.

The mean number of influenced activations over three 30-run Monte Carlo analyses is plotted in Fig. 7. Note that each of the sub-problems corresponding to one realization of the α values for all the nodes involved one run of Algorithm 1 with 1000 random samples (n). For the fourth case, the α values for the nodes are set deterministically in proportion to the node out-degree values. Clearly, it can be seen that the cases where all of the α values are either all small or all large fall short of the number of influenced activations for the case corresponding to the middle range of α values. The fourth case where the α values are proportional to the out-degree far outperforms the rest. This observation is in line with the discussions presented in Sect. 2.3.

5   A Centrality Metric Incorporating Intrinsic Activation

In this final section, we examine the influencer mining on networks with intrinsic and influenced nodal activations from a slightly different perspective. By collecting the various probabilities together and recognizing the recursive nature of influence spread on a social network, we arrive at a generalized PageRank-type spectral influence measure that was first presented in [22]. As demonstrated in [22], when considering activity on an OSN, this approach is a better measure of influence spread than a purely topological metric such as PageRank.

For a given node i, from Fig. 1, the total probability of activation $p_A^T(i)$ can be written as

$$p_A^T(i) = \alpha(i) + \left(1 - \prod_{j,(j,i)\in\mathcal{E}} \left(1 - \beta(i)w_{ij}p_A^T(j) \right) \right). \quad (2)$$

where $p_A^T(i)$ denotes the total probability of activation for node i (intrinsic and influenced). The parameter $\alpha(i)$ denotes probability of node i getting activated intrinsically. The quantity $\beta(i)w_{ij}$ denotes the probability of node j activating node i through influence as before.

Equation (2) summarizes the total activation probability for any given node in terms of the activation probabilities of the neighbors and whether the respective connecting edges are *live* or not. For the IC model, each activated node has a single shot probability of activating its neighbor and the activation along each edge is independent of the other edges. The second term in the large parentheses represents the probability of some neighboring node activating node i through influence. When we write out Eq. (2) for all the nodes, we will be dealing with a large system of coupled nonlinear equations, whose solution can be computationally expensive to obtain. By retaining the leading-order terms, we get a linearized version of Eq. (2). The goal of this approximation is to get the equation in a linear form so that we can use mature linear algebraic methods to quickly compute the set of influential nodes. This is valid to a large extent because, given the weighted-cascade version of the IC model that we are employing, the w_{ij} values will be small and we can neglect the higher-order terms to linearize Eq. (2) as follows.

$$p_A^T(i) = \alpha(i) + \beta(i) \sum_{j,(j,i)\in\mathcal{E}} w_{ij}p_A^T(j). \quad (3)$$

We set $\beta(i) = (1 - \alpha(i))$ as explained in Sect. 2.1 and extend Eq. (3) to the entire network with N nodes to obtain a matrix-vector equation as follows:

$$p_A^T = \alpha\mathbf{1} + ((I - \alpha)\,W)\,p_A^T. \quad (4)$$

In Eq. (4), p_A^T is a vector of size $N \times 1$, denoting respectively the total probability of activation for all the nodes on the network. I denotes the identity matrix of size $N \times N$. α denotes the diagonal matrix with entries corresponding to the intrinsic activation probability for all the nodes on the network; W denotes the sparse, stochastic weight matrix with entries given by the weights w_{ij} discussed earlier; and $\mathbf{1}$ is the all-ones vector of size $N \times 1$.

We can then express the total activation probabilities as

$$p_A^T = \mathbf{1}^T G; \quad G = (I - (I - \alpha)\,W)^{-1}\,\alpha, \quad (5)$$

where $\mathbf{1}^T$ provides for the column-sum of G. We also note that because the matrix W is a row stochastic matrix, the matrix G is also row stochastic.

Consider the quantity $C_A(i)$, specific to node i as defined below.

$$C_A(i) = \left(\sum_{j=1, i \neq j}^{N} G_{ji} \right). \tag{6}$$

$C_A(i)$ corresponds to the sum of the entries in column i of G with the exception of the corresponding diagonal term and represents the expected number of hosts activated by node i getting intrinsically activated and is a measure of influence. In our experiments with the PBlogs and the LFR-1000 graphs, discussed in Sect. 3.2, the α and W entries were randomized with entries drawn from the uniform distribution over $[0, 1]$ and W was converted to a row-stochastic matrix. We then compare the sets of top-k influencers identified by both the methods on two larger graphs in our dataset. The comparison is carried out with respect to two measures: (1) Jaccard similarity and (2) Rank-Biased Overlap (RBO). RBO considers ordering with higher weights given to matches that happen at the top [25]. These results are presented in Table 1 where we see good agreement between the sets of influential nodes obtained by both methods.

The behavior of the Jaccard index is not necessarily monotone as a larger number of influencers are considered. As we move away from the top influencing nodes (increasing k), we encounter many nodes that are of a similar influence. Since the centrality-based method is an approximation, the relative positions can change a lot and it is easily possible that going from $k = 30$ to $k = 50$, we may not get a proportional increase in the overlap between the two sets. Hence, it can result in non-monotone behavior. The RBO-based comparison can also exhibit a similar non-monotone behavior. However, this measure is known to be stable because of the weighting by the rank. The same is observed in our experimental results as well.

Thus, the proposed centrality metric, which includes the intrinsic activation mechanism, represents a computationally more viable alternative to the full-scale influence maximization framework. It retains the essence of the model and the influential nodes can be mined by solving a linear system involving a sparse matrix.

Table 1 Correlations, two ways, between the proposed approaches for the two inputs PBlogs and LFR1000 for different sizes of seed sets (10, 20, 30, and 50)

Correlation type	Input	$k = 10$	$k = 20$	$k = 30$	$k = 50$
Jaccard	PBlogs	0.538	0.818	0.875	0.818
RBO	PBlogs	0.817	0.846	0.851	0.868
Jaccard	LFR1000	0.818	0.905	0.765	0.818
RBO	LFR1000	0.979	0.963	0.947	0.937

Closer the metric to one, the better

6 Conclusions

In this work, we introduce the notion of nodes in a complex network getting activated by two mechanisms: intrinsically and through influence as commonly observed in online social networks. Using a modified version of the influence maximization algorithm and working with a suitable influence spread objective function, we show how it is possible to identify influential users whose intrinsic content spreads maximally through influence. We also sketch a short proof on the submodularity of the modified influence function, allowing for approximation guarantees on the algorithm. We utilized several synthetic and real-world datasets to examine various aspects of the proposed activation model. We also explain why some assignments of the intrinsic activation probability (α values) to the various nodes can result in much higher activations than other assignments, which is also demonstrated on a Twitter graph. We finally derive a novel centrality metric from the activation model that can provide for a computationally faster and accurate method to identify influential users on a social network where activations can be intrinsic or influenced.

Building on this work, we are exploring multiple facets of this problem in our ongoing research including the exploration of how a social network can be successful in the long run by balancing the two modes of activation discussed here. To achieve this objective, we are considering development of variants of other interaction models with intrinsic activation as well extending the notion of intrinsic activation to more fine-grained user behavior. We are also extending these methods to other complex systems, such as for attack modeling in cyber networks [20].

Acknowledgements This research was supported in part by the High Performance Data Analytics Program (HPDA) and in part by the Control of Complex Systems Initiative (CCSI) at the Pacific Northwest National Laboratory (PNNL). HPDA is a collaboration led by Pacific Northwest National Laboratory (PNNL) with partners Mississippi State University, University of Washington, and Georgia Institute of Technology. CCSI is a Laboratory Directed Research and Development (LDRD) program at the PNNL. PNNL is operated by Battelle for the U.S. Department of Energy under Contract DE-AC05-76RL01830.

References

1. Adamic, L.A., Glance, N.: The political blogosphere and the 2004 us election: divided they blog. In: Proceedings of the 3rd International Workshop on Link Discovery, pp. 36–43. ACM, New York (2005)
2. Bastian, M., Heymann, S., Jacomy, M.: Gephi: an open source software for exploring and manipulating networks (2009). www.gephi.org
3. Borgs, C., Brautbar, M., Chayes, J., Lucier, B.: Maximizing social influence in nearly optimal time. In: SODA '14 Proceedings of the Twenty-Fifth Annual ACM-SIAM Symposium on Discrete Algorithms, pp. 946–957. SIAM, Philadelphia (2014)

4. Chen, W., Collins, A., Cummings, R., Ke, T., Liu, Z., Rincon, D., Sun, X., Wang, Y., Wei, W., Yuan, Y.: Influence maximization in social networks when negative opinions may emerge and propagate. In: SIAM Data Mining, pp. 379–390 (2011)
5. Cohen, E., Delling, D., Pajor, T., Werneck, R.F.: Sketch-based influence maximization and computation: scaling up with guarantees. In: CIKM '14 International Conference on Conference on Information and Knowledge Management, pp. 629–638. ACM, New York (2014)
6. Farajtabar, M., Du, N., Gomez-Rodriguez, M., Valera, I., Zha, H., Song, L.: Shaping social activity by incentivizing users. In: Advances in Neural Information Processing Systems, pp. 2474–2482 (2014)
7. Friedkin, N.E., Johnsen, E.C.: Social influence networks and opinion change. Adv. Group Process. **16**(1), 1–29 (1999)
8. Gionis, A., Terzi, E., Tsaparas, P.: Opinion maximization in social networks. In: SIAM Data Mining Conference, pp. 387–395. SIAM, Philadelphia (2013)
9. Halappanavar, M., Sathanur, A., Nandi, A.: Accelerating the mining of influential nodes in complex networks through community detection. In: CF'16 Proceedings of the 13th ACM International Conference on Computing Frontiers, Como, May 16–18, 2016, pp. 64–71 (2016)
10. Kempe, D., Kleinberg, J., Tardos, E.: Maximizing the spread of influence through a social network. In: Proceedings of ACM SIGKDD, pp. 137–146. ACM, New York (2003). https://doi.org/10.1145/956750.956769
11. Kempe, D., Kleinberg, J., Tardos, É.: Influential nodes in a diffusion model for social networks. In: Automata, Languages and Programming, pp. 1127–1138. Springer, Berlin (2005)
12. Krause, A., Golovin, D.: Submodular function maximization. In: Tractability: Practical Approaches to Hard Problems, vol. 3(19), p. 8. Cambridge University Press, Cambridge (2012)
13. Lancichinetti, A., Fortunato, S.: Benchmarks for testing community detection algorithms on directed and weighted graphs with overlapping communities. Phys. Rev. E **80**(1), 016118 (2009)
14. Leskovec, J., Krause, A., Guestrin, C., Faloutsos, C., VanBriesen, J., Glance, N.: Cost-effective outbreak detection in networks. In: Proceedings of ACM SIGKDD, pp. 420–429, ACM, New York (2007)
15. Li, Y., Chen, W., Wang, Y., Zhang, Z.L.: Influence diffusion dynamics and influence maximization in social networks with friend and foe relationships. In: Proceedings of the Sixth ACM International Conference on Web Search and Data Mining, pp. 657–666. ACM, New York (2013)
16. Mirzasoleimanm, B., Badanidiyuru, A., Karbasi, A., Vondrak, J., Krause, A.: Lazier than lazy greedy. In: Twenty-Ninth AAAI Conference on Artificial Intelligence (2015)
17. Myers, S.A., Zhu, C., Leskovec, J.: Information diffusion and external influence in networks. In: ACM SIGKDD, pp. 33–41. ACM, New York (2012)
18. Quach, T.T., Wendt, J.D.: A diffusion model for maximizing influence spread in large networks. In: Social Informatics: 8th International Conference, SocInfo 2016, Bellevue, November 11–14, 2016. Proceedings, Part I, pp. 110–124. Springer International Publishing, Berlin (2016)
19. Saito, K., Nakano, R., Kimura, M.: Prediction of information diffusion probabilities for independent cascade model. In: International Conference on Knowledge-Based and Intelligent Information and Engineering Systems, pp. 67–75. Springer, Berlin (2008)
20. Sathanur, A.V., Haglin, D.J.: A novel centrality measure for network-wide cyber vulnerability assessment. In: IEEE Symposium on Technologies for Homeland Security (HST), 2016, pp. 1–5. IEEE, Washington (2016)
21. Sathanur, A.V., Halappanavar, M.: Influence maximization on complex networks with intrinsic nodal activation. In: Social Informatics: 8th International Conference, SocInfo 2016, Bellevue, November 11–14, 2016. Proceedings, Part II. Springer International Publishing, Berlin (2016)
22. Sathanur, A.V., Jandhyala, V., Xing, C.: Physense: scalable sociological interaction models for influence estimation on online social networks. In: IEEE International Conference on Intelligence and Security Informatics, pp. 358–363. IEEE, Washington (2013)

23. Srivastava, A., Chelmis, C., Prasanna, V.K.: Influence in social networks: a unified model? In: 2014 IEEE/ACM International Conference on Advances in Social Networks Analysis and Mining, pp. 451–454. IEEE, Washington (2014)
24. The Twitter public API. https://dev.twitter.com/rest/public
25. Webber, W., Moffat, A., Zobel, J.: A similarity measure for indefinite rankings. ACM Trans. Inf. Syst. **28**(4), 20 (2010)
26. Zhang, H., Mishra, S., Thai, M.: Recent advances in information diffusion and influence maximization in complex social networks. In: Opportunistic Mobile Social Networks, p. 37. CRC Press, Boca Raton (2014)

Influence and Extension of the Spiral of Silence in Social Networks: A Data-Driven Approach

Yingbo Zhu, Zhenhua Huang, Zhenyu Wang, Linfeng Luo, and Shuang Wu

Abstract The Spiral of Silence has been studied widely in the traditional propagation field. However, to our best knowledge, no one has clearly verified the Spiral of Silence in social networks based on the real information diffusion data. In this paper, four factors including width, depth, message sentiment, and modularity of information diffusion trees are analyzed to verify the applicability of the theory. Disparities between majority and minority are found to different extents in various topics. Based on Spiral of Silence, polarity prediction of users' review without considering the semantic meaning of content is proposed and discovered. The results indicate that opinions of people in propagation are impacted by social environment and their friends. The Anti-Spiral of Silence, an extension of Spiral of Silence, has been found to play a significant role in leading rational public opinion and revealing truth in social networks. Our works of both Spiral of Silence and Anti-Spiral of Silence will enrich research results on the study and application of propagation effects.

Keywords Spiral of Silence · Information propagation · Prediction of public opinion · Anti-Spiral of Silence

1 Introduction

With rapid development of Web 2.0, social networks such as Facebook, Twitter, LinkedIn, and Sina Weibo provide convenient platforms where people express, share, and exchange their opinions and ideas. Social networks give people opportunities to express themselves in a more straightforward way. Sina Weibo, a

Y. Zhu
Tianyi Music Culture & Technology Co. Ltd. China, Guangzhou, China
e-mail: zhuyb@118100.cn

Z. Huang · Z. Wang (✉) · L. Luo · S. Wu
South China University of Technology, Guangzhou, China
e-mail: wangzy@scut.edu.cn

© Springer International Publishing AG, part of Springer Nature 2018   143
M. Kaya et al. (eds.), *Social Network Based Big Data Analysis and Applications*,
Lecture Notes in Social Networks, https://doi.org/10.1007/978-3-319-78196-9_7

twitter-like microblog platform, one of the most popular online social networks in China, has about 297 million active users per month in the third quarter of 2016 [1].

People show their attitudes about hot events in social networks, and the public opinions related to certain events evolution as time goes. People usually read reviews of others and make their comments after that. A special phenomenon of Spiral of Silence (SOS) may be formed in the process of information propagation. The Spiral of Silence, proposed by Elisabeth Noelle-Neumann in 1973 [2], is one of the most important theories in the traditional propagation field. The theory stipulates that individuals have a fear of isolation, resulting in keeping silence when they find their opinions conflict with the majority. Due to fear of isolation, the minority people remain silent instead of expressing opinions. As a result, with time elapsing, the disparity between majority and minority opinions increases and causes a spiral. The phenomenon is called the Spiral of Silence. The Spiral of Silence, a representative theory of communication theory, enjoys a high reputation among research on public opinion studies.

The Spiral of Silence theory has been studied widely in the traditional dissemination field [3]. However, with changes of social media environments, scholars have raised doubt on whether the Spiral of Silence theory still works in social networks [4]. Most of the researches were based on simulation experiments or questionnaires, but no one before had studied the Spiral of Silence theory based on the real information propagation datasets in social networks. Investigating the applicability of the Spiral of Silence theory in social networks is still an open issue with much challenge. Furthermore, understanding the Spiral of Silence theory in social networks is also useful for advertising assessment and public opinion monitoring. Therefore, applicability verification of the Spiral of Silence theory in social networks deserves deep investigation.

The applicability of Spiral of Silence in social networks is verified in our prior works on ASONAM 2016 [5]. To testify it, information dissemination trees (also called cascades) about how events propagate among users in Sina Weibo are reconstructed. Disparities of four factors including width, depth, sentiment, and modularity are analyzed to verify the applicability of the Spiral of Silence in social networks. In this paper, we extend our research and propose some new scopes. First, the old method that divided the whole life cycles into twenty slices ignored the characteristics of user behavior. Time interval of user retweet behavior follows a power law distribution. Several users retweet microblogs may extend the life cycle into months. If we divide the life cycle of a cascade evenly, there will be a huge unbalance in different time slices, shown in Fig. 1. So we promote the dividing strategy and testify whether the Spiral of Silence still works in this circumstance. And the dataset of information diffusion trees is enriched. In the new dataset, the categories of topics are labeled by Sina Weibo and there are 44 topics in total.

The theory of Spiral of Silence is significant in studying public opinions, but how to apply the theory into polarity prediction of uses still remains open. Since opinions of people are impacted by majority, based on Spiral of Silence, a problem is proposed that can we predict sentiment polarity of people toward an event? Does

(a)   (b)

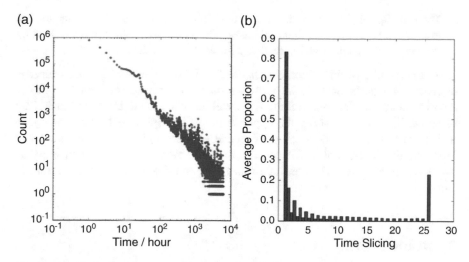

Fig. 1 (**a**) Distribution of user retweet time. (**b**) Proportion of reviews in each slicing time, the blue ones represents new strategy

social relation influence people's attitudes and to what extent it impacts on people? We defined the problem of predicting polarity of users' opinion without considering the semantic meaning of content, achieving a precision of 84.7% at most. The results have proven that sentiment of people is influenced by social environment. Sentiment of parent nodes is shown to have significant impacts on user opinions especially.

People in social networks can have their own opinions and not always keep silent. Message diffusion patterns in some categories are found not corresponding to the Spiral of Silence very well, which indicates there may be other diffusion mechanism besides the theory in social networks. Some scholars proposed the Anti-Spiral of Silence [6]. Does Anti-Spiral of Silence exist in social networks at the same time? What kinds of roles Anti-Spiral of Silence play in evolution of public opinions? We have testified that the Anti-Spiral of Silence really exists in real cases. Furthermore, its effects in rational thinking and truth in public events are also revealed.

The highlights of this paper include

1. Based on the real datasets of information propagation, we have verified the applicability of the Spiral of Silence theory in social networks by taking four factors into account, including width, depth, message sentiment, and modularity. Compared with existing research, we promote the life cycle dividing strategy by considering characteristics of user behavior. Under different topics of messages, Video, Movie, Mood, and Food tweets are found to have high applicability of the theory, while Society has weak applicability.
2. Based on the Spiral of Silence, a problem of predicting polarity of users' reviews without considering semantic meaning is proposed. Experimental results have proven that it was predictable to some extents and opinions of people are influenced by social environment and friendship relations.

3. The Anti-Spiral of Silence, an extension of Spiral of Silence, has been testified to exist in social networks using real cases. The significant role it plays in leading rational thinking and reveling truth is also described in detail.

The remainder of this paper is organized as follows. Section 2 is an overview of related works. In Sect. 3, we will introduce the preliminaries and definitions involved in the verification. In Sect. 4, we will introduce the datasets and main idea of the verification, and verify the applicability of the Spiral of Silence theory. In Sect. 5, we will discuss the problem of predicting users' opinion. In Sect. 6, we will introduce the Anti-Spiral of Silence and explain how the effects influence events evolution. Finally, we will conclude this paper and discuss directions of further work.

2 Related Work

The Spiral of Silence theory has attracted great interest of research since it was proposed, and it also has been studied extensively in both traditional dissemination field and network environment. Moy et al. [7] have studied the Spiral of Silence theory and found that isolation is a motivating factor for individuals to speak out or not. The nature of the relationship is another factor that affects the Spiral of Silence theory [8]. Gearhart et al. [9] indicated that one's attitude is also an important variable that influences whether individuals voice opinions or keep silent. Matthes et al. [10] applied the main idea of the theory to discuss how the public opinion influenced deinstitutionalization of an event. These research results are very helpful to improve the understanding and application of the Spiral of Silence in the traditional dissemination field.

In the network environment, Zuercher's research [11] has shown that due to the lack of contextual social cues, the minority were willing to express their opinions more freely without being isolated. Lee et al. [12] indicated that ideology played a key role in an individual's online participation. Yue et al. [13] have studied the Spiral of Silence theory by taking into account four factors in three different networks through simulation experiments. Zerback et al. [14] discovered the Spiral of Silence theory by linking it to exemplification research. Spiral of Silence is found to exist on social networks from large investigation on people's exposure to information on important political issues and their willingness to discuss these issues with those around them according to Hampton et al. [15]. Lee et al. [12] have surveyed 118 journalists about their behaviors on Twitter with regard to two controversial issues in South Korea based on the theory of the SOS. The journalists' ideology was found to be a significant factor in expressing their opinions about controversial issues on Twitter. But none of them used real information propagation data to analyze existence and influence of Spiral of Silence.

Predicting click behavior on Facebook was discovered by Pang et al. [17] based on the Spiral of Silence. Tan et al. [16] have found that users who are somehow connected are more likely to hold similar opinions. Therefore, relationship information can complement what we can extract about a user's opinions from their utterances. They demonstrated that user-level sentiment analysis can be significantly improved by incorporating social information. The application of Spiral of Silence in prediction and how the Spiral of Silence works in the propagation process in real data still need to be deep explored.

The Anti-Spiral of Silence was found effective in social networks with Spiral of Silence at the same time [5, 18]. Wang et al. believed that there were a considerable number of rational users in the network. These users had a strong sense of responsibility and justice to society and were courageous to express their views. It had positive effects on disclosure of the truth and supervision of social behavior [18].

Our research is different from existing works; applicability of the Spiral of Silence in social networks is verified through the real-world information propagation data. Through investigating applicability of tweets with different categories, we find that propagation in various topics has applicability of the Spiral of Silence to different extents. Based on the Spiral of Silence, a problem of predicting sentiment of users' reviews without considering semantic meaning is proposed. Social environment and friendship relation are found to have important impacts on user's opinion. And we give two examples from real data to show existence of Anti-Spiral of Silence and explain its functions in public opinion. The Anti-Spiral of Silence can be regarded as an extension of Spiral of Silence and is helpful for people to think rationally and deeply about the truth.

3 Preliminaries

3.1 Information Propagation in Social Networks

Relationships between individuals play a critical role in the information propagation process in Sina Weibo. Resharing functionality allows users to retweet or repost messages and facilitate propagation of messages. Therefore, many information dissemination trees or cascades are formed. An information propagation tree is shown in Fig. 2.

Fig. 2 An example of information dissemination tree

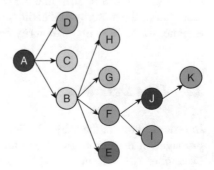

Fig. 3 A real example of the
theory in Sina Weibo

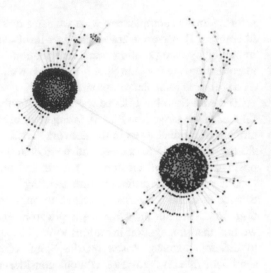

As shown in Fig. 2, each node in the tree represents a tweet, Node A is the
root of the tree where original tweets are released in Sina Weibo. The edge with
an arrow represents the direction of propagation. Each node has its depth and
contains information of review (comment), user ID, and retweet time. In this simple
example, the information propagation tree has been transmitted for ten times and
finally formed a structure with four layers. We applied the breadth-first approach
to obtain all the complete propagation trajectories of a message and constructed an
information dissemination tree about the message.

As shown in Fig. 3, it is a real information dissemination tree in Sina Weibo. The
reviews on the tree are classified into positive and negative categories by LingPipe
[19], which is an open-source toolkit for processing text using computational
linguistics to analyze message sentiment. Then we use Gephi [20], an open-
source graph visualization and manipulation software, to visualize the information
propagation tree, and we mark the nodes of majority and minority with red and
green colors, respectively. After applying the layout algorithm of Yifan Hu [21], we
can intuitively know that the majority has dominated the public opinion from Fig. 3.
Moreover, nodes in the graph are separated into two parts which are shown in Fig. 4.
From Fig. 4, it can be seen that the connections within the majority are far denser
than that of the minority, indicating nodes in the majority are more likely to gather
together while those in the minority do not.

3.2 Data Description

In order to prevent the subjective impact, we randomly collect 11,235 information
propagation trees from the "hot event" in Sina Weibo between March 2013 and
December 2016. There are 5200 diffusion trees with forwarding tweets over 50,

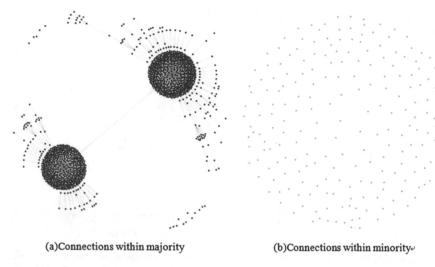

(a)Connections within majority   (b)Connections within minority

Fig. 4 Connections within two opinions

Fig. 5 The proportion of
different categories

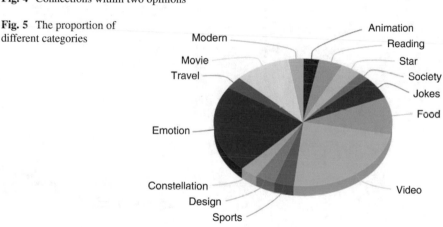

which are reserved to analyze the Spiral of Silence. Since there are so many topics
in the dataset, we only present categories with a number of diffusion trees more than
2% in Fig. 5. The size and depth distribution of the trees are shown in Figs. 6 and 7.

The strategy of evenly dividing the life cycle ignores the characteristics of
forwarding behavior. User forwarding time (time interval with the original message
publishing time) follows a power-law distribution as shown in Fig. 1. A small
number of users with high forwarding time may extend the life cycle into months.
But the distribution does not have "long tails" effects. If a "long tails" exists, it
means the longer forwarding time the fewer users. A reasonable explanation for the
phenomenon is that forwarding behavior has a cascading effect. Some users may be
involved in forwarding messages long after the messages are published. But they
can lead more people to forward the messages in a short period. And this reflects

Fig. 6 Cascade size
distribution

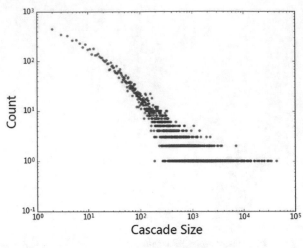

Fig. 7 Cascade depth
distribution

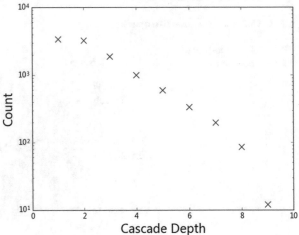

existence of Cascades Recur. In the method of diving the life cycle, the average of the first forecast time is more than 50 hours after release of original message. In this way, the first slice contains 83.4% microblogs. While in the first hour it contains only 16.2% microblogs, reducing extreme imbalance of data and making the analysis more reasonable.

3.3 Tweet Sentiment Analysis

In order to analyze users' opinion sentiment, we consider the message sentiment analysis as a binary classification problem, we classify all comment messages into positive and negative categories by LingPipe. We train our model based on a language model which is provided by LingPipe with a large amount of corpus which contain 5350 positive and 5318 negative messages. The message sentiment value $S(i)$ of the information propagation tree at the i-th time point is calculated by the following formulation:

$$S(i) = \sum_{k=1}^{i} N_S^k \times \frac{\sum_{k=1}^{i} N_k}{N_A} \tag{1}$$

In Eq. (1), NA is the sum of the message in the information propagation tree. Ni and NSi represent the number of messages and the number of positive or negative messages at the ith time point, respectively.

By the above definition, sentiment category of each review and sentiment strength of each category at each time point are calculated.

3.4 Width and Depth

As is mentioned in Sect. 3, each node has its depth. In the process of propagation, the structure of the information dissemination tree is changing as time goes on. In this paper, with reference to the definition of width of a binary tree, we define the width of an information propagation tree at each time point as the maximum number of positive or negative nodes in all layers of the tree at each time point. The average depth of positive or negative nodes in the information propagation tree at each time point is analyzed. The width $W(i)$ and depth $D(i)$ of an information propagation tree at the i-th time point are calculated by the following formulations:

$$W(i) = \mathrm{MAX}\left(w_S^{kl}\right), \quad k = \{1, 2 \ldots, i\}, \quad l = \{1, 2, \ldots n\} \tag{2}$$

$$D(i) = \frac{\sum_{k=1}^{i} d_S^k}{\sum_{k=1}^{i} N_k} \tag{3}$$

In Eq. (2), $wSil$ is the number of positive or negative nodes in the l-th layer of the information propagation tree at i-th time point. In Eq. (3), dSi and Ni respectively represent the depth of positive or negative nodes and the number of nodes of the information propagation tree at the ith time point.

3.5 Modularity

Modularity, proposed by Mark Newman [22], was originally applied to evaluate quality of community detection algorithms. Groups with high modularity have dense

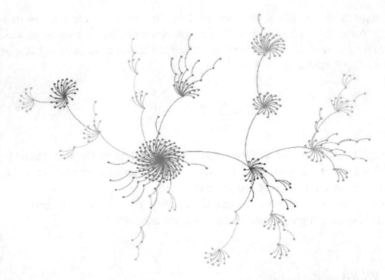

Fig. 8 An example of a diffusion tree with viral structure. The id is 3931505087003828, modularity is 0.835

Fig. 9 An example of a
diffusion tree with broadcast
structure. The id is
3931929482871856,
modularity value is 0.039

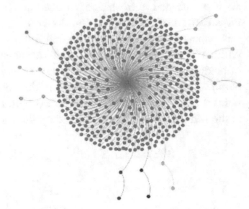

connections between the nodes within modules but sparse connections between nodes in different modules. After detecting communities in propagation trees, the modularity of the partitions can distinguish different types of structure. The diffusion tree with high structural virality [23] has higher modularity value and the messages in these trees propagate like a virus from person to person as shown in Fig. 8. While as Fig. 9 shows, the diffusion trees with small modularity value are star-like in which most paths are shallow. We apply the algorithm in Blondel et al. [24] to detect communities in cascades and calculate modularity using the following formulation, wherein $A_{i,j}$ is actual edge between node i and j, k_i is the degree of node i, and the value δ depends on whether nodes i and j are in the same community.

$$Q = \frac{1}{2m} \left[A_{i,j} - \frac{k_i k_j}{2m} \right] \delta\left(c_i, c_j\right) \tag{4}$$

4   Verification of the Spiral of Silence Theory in Social Networks

In this section, we will introduce the main idea to verify the Spiral of Silence's applicability in social networks. The theory indicates that humans are fearful of being isolated. People would like to keep silent instead of voicing opinions when their opinions are not in majority. Thus, with time elapsing, the disparity between majority and minority opinions will become larger and larger. We focus on analyzing the disparity between majority and minority opinions. Four factors are taken into account in experiments including the propagation width, depth, message sentiment, and modularity based on the information propagation datasets collected from Sina Weibo.

First, nodes are classified into positive and negative categories and two information propagation trees are constructed. The category that has larger quantity is regarded as the majority opinion, and the other one as the minority opinion. Then difference of the above mentioned four factors is calculated respectively between majority and minority opinions at each time point. Finally, we analyze the results and explain the reason why the spiral of silence theory still works in social networks.

4.1   Width Analysis of Information Propagation

In this section, we focus on the width disparity of majority and minority opinions. We conduct the experiments concerning the tendency of width disparity as time elapses and verify the applicability of the Spiral of Silence theory from the propagation width aspect. As shown in Fig. 10, as time goes on, the width disparity of majority and minority opinions becomes larger. Generally, when individuals want to repost a tweet, most of them will first look up the comments of the tweet, and then there will be a high probability for them to repost the tweet after they realize their opinions are similar to the public opinion. On the contrary, once they find their opinions are opposite to the public opinion, most of them would like to keep silent instead of voicing. Thus, the result indicates that more people prefer to repost the tweet which represents majority opinion rather than the one represents minority opinion.

This result can also be explained by the Spiral of Silence theory: In order not to lose popularity or be isolated, individuals constantly observe the public opinion, individuals behave and express themselves accordingly in public, and they estimate the climate of public opinion by different ways, such as public media. They then determine whether they are in the majority or minority; if they feel that they belong to the majority, they would like to express their opinions. However, they tend to remain silent if they feel they are the minority.

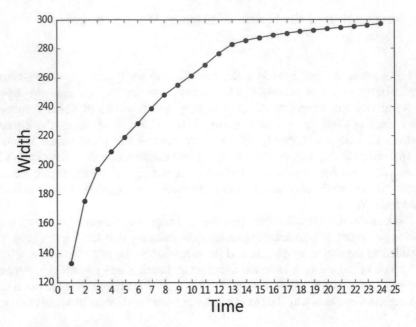

Fig. 10 Distribution of width disparity

4.2 Depth Analysis of Information Propagation

In this section, we will verify the applicability of the spiral of silence theory by investigating the depth disparity of majority and minority opinions. The distribution of depth disparity is shown in Fig. 11.

Interestingly, the distribution of depth disparity is similar to the distribution of width disparity. Although the propagation width and depth have the similar distribution, the meanings behind them are quite different. When individuals want to express their opinions, most of them would like to repost the tweet which has a larger depth because they consider the behavior of reposting a tweet as a process of energy transfer, and they reckon that the energy will be stronger with their reposts. And majority opinion is considered as energy and individuals whose opinions are similar to the majority opinion would like to transfer the energy by reposting the tweet, finally the depth disparity between majority and minority opinions becomes larger.

4.3 Message Sentiment Analysis of Information Propagation

In this section, we focus on the disparity between the sentiment message value of majority and minority opinions. We compute the message sentiment value of the two opinions at each time point by Eq. (1). The distribution of the message sentiment disparity is shown in Fig. 12.

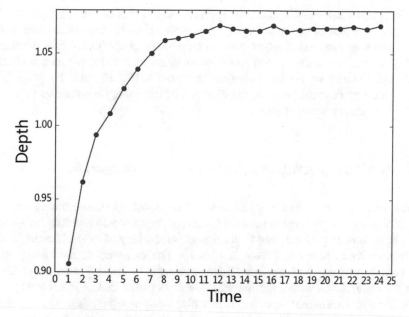

Fig. 11 Distribution of depth disparity

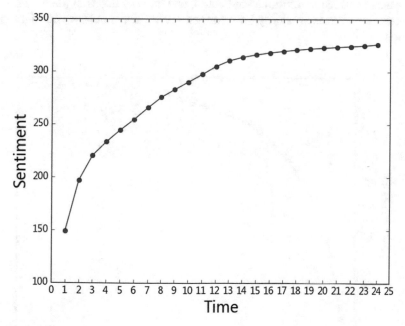

Fig. 12 Distribution of sentiment disparity

The result confirms that the message sentiment value of the majority opinion is larger than that of the minority opinion with time elapsing which means that the majority opinion has stronger emotion than minority opinion. When individuals repost a tweet, they not only express their own opinions but also enhance the energy of the opinion they supported. Therefore, as shown in Fig. 12, with the information propagation process carrying on, the disparity of the energy between majority and minority opinions becomes larger.

4.4 Modularity Analysis of Information Propagation

In this section, we verify the applicability of the spiral of silence theory from the modularity aspect. We first extract the graphs of majority and minority opinions at each time point respectively and then calculate modularity of them. The distribution of modularity disparity is shown in Fig. 13. The disparity of modularity value between the majority and minority opinions presents an upward trend which means that the connections within the majority are denser than that of the minority with time elapsing. Due to the fear of isolation which is stipulated by the Spiral of Silence theory, the minority would not like to expose their opinions to the public, so it cannot form communities within the minority, it also can be confirmed from the real example which is mentioned in Sect. 3, and most of nodes in the minority are isolated. However, the majority have formed several communities with tight inner relations.

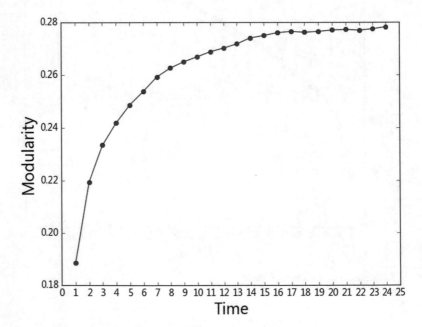

Fig. 13 Distribution of modularity disparity

4.5 Theory's Applicability Analysis of Different Categories

In this section, six categories are chosen to analyze the Spiral of Silence in various topics. The results are shown in Fig. 15. It is found that tendencies of disparity in Video topic rank first in the three aspects—width, modularity, and sentiment. In message spreading of category Video, the Spiral of Silence has significant effects. While in depth disparity, Video ranks the last. It indicates that users tend to choose forward from root users in Video propagation. Although as time goes, the number of the forwarding users are increasing, but the depth disparity remains stable or even decreases a little. The tendency patterns are similar among Movie, Food, Mood, and Reading in width, modularity, and sentiment. It shows that the theory has the strong applicability in these categories, which means that the majority opinion has dominated the public sphere. And there seems to be a sweet point of 12th slice where the disparity keeps unchanged after the time point. For the movie category, its disparity is high and tendencies are almost of the same pattern, which means that they have the strong applicability of the theory. However, for Society topic, the tendency pattern in modularity and depth is far different from those of others. It implies that the structure of diffusion trees in society topic is more viral. The debates in society are more intense and users are more likely to forward the microblog retweeted by their friends. The phenomenon causes the formation of many deep paths and many secondary outbreaks as Fig. 14. And increasing patterns in width and sentiment are almost the same. Before the 12th hour, the width and sentiment are growing very fast and almost stop growing after the 12th hour. Society topic has different tendency in modularity and depth aspects but has similar tendency of width and sentiment aspects, this may be because both majority

Fig. 14 Example of how an event with society topic propagates

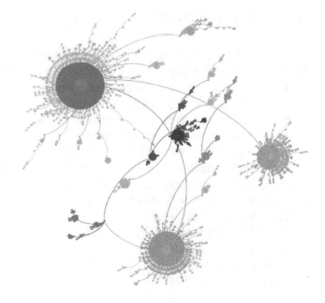

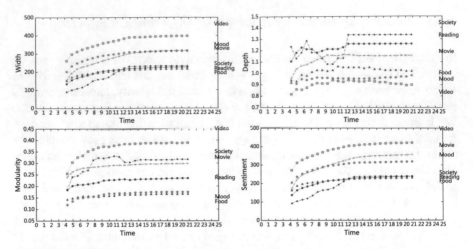

Fig. 15 Experiment results of the six categories

and minority opinions have tight relations between themselves separately and make public opinions fluctuate as time goes. While disparity of modularity in mood and food is very small because the most nodes are directly linked with the root node. The modularity of these diffusion trees themselves is very small and there is a relatively smaller disparity between majority and minority.

5 Polarity Prediction of People's Opinion

5.1 Problem Statement

The Spiral of Silence has described a phenomenon that people instinctively seek for support when they express their opinions and fear for isolation. The silence of opinions in one part leads to an increase of opinions in another part. In this way, it causes one part with more powerful voice and the other part becomes more and more silent, forming a spiral development process. Since people's opinion is related to social environment. Can we use these information to predict people's opinion toward an event? We extract rich features potential to be related to user reviews and train classifiers to make prediction. We regard the problem as a binary classification problem, predicting whether a user will hold a positive or negative opinion. The sentiment of review context is preprocessed and used to label data.

The problem is defined below:

$T = (V, E)$ represents a diffusion tree. V represents all users involve in the diffusion. E indicates the propagation links. When a user u retweets and comments the message in time t, predicting opinion polarity of the user, the current information propagation tree is noted as T^t.

5.2 Feature Extracted

Now we introduce features related to the problem.

User features: Persons with a huge influence may have a higher possibility to express their true opinions. To measure user influence, a number of fans are used. And features of gender, user rank, and VIP member rank are also taken into consideration.

Temporal Features: The time intervals of retweet reflect the speed of the propagation that may affect people's opinion.

Parent Features: Sentiment of the parent node may also have an impact on users' opinions.

Other Features: Whether a review contains a "@" or "#" and length of the review context.

Silence Features: Depth and width of the current diffusion tree. The message sentiment and modularity of two parts that hold different opinions.

5.3 Experiments and Analysis

Nodes in diffusion trees are sampled randomly and features related to these nodes are extracted to construct training data. The nodes that do not have to review content or just retweet are ignored. The sentiment of the content is regarded as a class label. A variety of classifiers are employed to train model and make prediction. Due to the Spiral of Silence, people who hold opinions of positive and negative is imbalance. We apply the method of sampling to balance the dataset of positive and negative samples [25]. Then classifiers are trained based on the balanced data. The sampled data includes 34,825 positive and negative samples separately.

The results are as Table 1 when all features are used. The Random Forest performed best among the classifiers. The highest precision is 84.7% while the baseline of random guess is 50.02%. The results show that social environment has important impacts on people's opinion. To discover whether we could predict a user's opinion before they make comment, the F-measure is 66.7%, still much higher that baseline. It indicates that social environment has high influence on people's willingness to show their opinion. To discover whether opinions of users are related to the sentiment of parent node, we remove the feature of "Parent Features" and make comparison. The F-measure has decreased by 5.3%. The parent node usually is opinion leader among its child nodes or a friend of the child node. So the parent node has a strong impact on the sentiment of the child node. It shows that user opinion is influenced by social relationship, which is corresponding to research before [16]. And people are more likely to show their opinion if they hold the same emotion with their parent nodes, and the phenomenon intensifies the effects of the Spiral of Silence.

Table 1 Results in
prediction

	SVM	C4.5	Random forest
Precision	0.839	0.838	0.847
Recall	0.830	0.837	0.846
F1	0.830	0.837	0.846
ROC	0.833	0.876	0.928

6 The Anti-Spiral of Silence

The Internet environment brings expression space with more freedom and violates the conditions of the Spiral of Silence theory to a certain extent. Some famous events that occurred in the current Internet may show opposite characteristics with Spiral of Silence, and we have found that the Spiral of Silence is no longer the golden rule. The Anti-Spiral of Silence develops the Spiral of Silence theory in the new era of social networks. The Anti-Spiral of Silence is a kind of new theory of propagation and can be regarded as an extension of the Spiral of Silence. According to the theory, people are initiative agents that can self-think and self-analyze, not blind to conformity and convergence. The audiences are courageous to express their opinions or support a minority opinion. If the minority opinions are accepted by more and more rational users, it becomes well-matched in strength with the majority opinions or even beyond advantage part of the views by spiral evolution. This situation is manifested as the Anti-Spiral of Silence.

Through analysis of the whole datasets, the Spiral of Silence testified exists in social networks. However, disparity varying in different categories may indicate that Anti-Spiral of Silence may exist in some cases. We choose two famous examples in Sina Weibo to verify existence and demonstrate effects of the Anti-Spiral of Silence.

The first cascade is about the event "Yixiao Luo" shown in Fig. 16, a famous event in social networks. Luo was a little girl who got leukemia. A passage about her was published by a company and was forwarded many times and got huge funding in social networks. At first, people showed pity and compassion for the girl. At this time, positive opinion was the majority and disparity between majority and minority increased with time. But the passage was suspected of using love marketing by some users. It led to growth of negative opinion. So there is a sharp decrease after several slices in width. It indicates that the newly published reviews of negative sentiment are much more than those of positive sentiment. And at that time, the people who hold the new negative opinion were more likely to agree and retweet from new reviews but not root review, causing a sharp decrease in disparity of depth. After some people had released the information that the medical fee was not that high and the girl's father even had three houses. Denounced and angry voice became more and more powerful. At this time, width of negative reviews even exceeds that of positive reviews and the disparity even increased as time goes. As the truth was constantly being revealed, most reviews condemned her father and the company. And debate was very intensive in social networks. Under the pressure of public

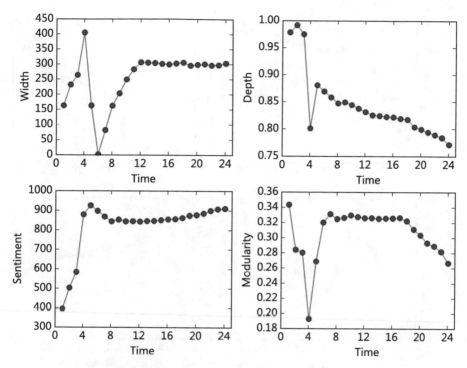

Fig. 16 Disparity of public opinion in event "Yixiao Luo"

opinion, Luo's father and the company had to speak in a statement to clarify the facts. At last, involvement of the State Department and the Tencent Incorporate has given back all the donations.

The event in Fig. 17 is about "Divorce of Baoqiang Wang." Another possible reason leading to increasing strength of the weak side is aspects that people care about have changed. Wang is a famous start in china. He announced his divorce in Weibo and aroused wide attention. In the event of "Baoqiang Divorce," most users cared and supported Baoqiang. A user called Mango Song blamed Wang for his negligence of the family. This message had aroused a heat discussion. Some people began to think about the event rationally while the others criticized the person, increasing a large amount of negative reviews in short time. And width of negative reviews exceeds that of positive reviews at third time point. However, message sentiment of both positive and negative reviews increased with the time.

There are a variety of rational and irrational reviews in social networks. When the irrational opinion occupies high public opinions in the new media environment, many unsuspecting users will change their opinion or keep silence. The truth can be hidden under this situation. But some opinion leaders are a group of people who insist on truth or rational points of view. They can analyze public opinions rationally and guide users out of passion trap to make correct judgment. Even if a rational point of view is weak in quantity at first, they do not keep silent and lead more people to

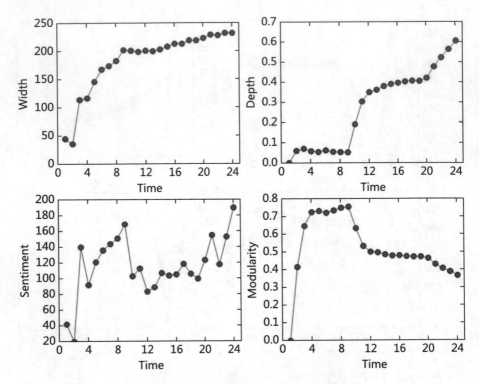

Fig. 17 Disparity of public opinion in event "Divorce of Baoqiang Wang"

speak out, causing failure of Spiral of Silence. With the development of the incident and disclosure of truth, more users come back to rational thinking, resulting in more people adhering to the correct opinion. The rational points of view gradually grow from weak to strong, and irrational points of view are suppressed.

In the above cases, the theory of Anti-Spiral of Silence is shown to play a positive role in guiding the correct public opinion and values. In the event of "Yixiao Luo," some people stuck to their own views and questioned the incident. They led users to think more rationally and put pressure on Luo's father and company. In the event of "Divorce of Baoqiang Wang," the Anti-Spiral of Silence can change aspects of thinking and help people debate more. It guides the direction of public opinion and prevents the truth from being covered up.

7 Conclusions and Future Work

In this chapter, we have verified the applicability of the Spiral of Silence theory from four aspects on Sina Weibo by data-driven experiments. Compared with our earlier works, we have taken characteristics of user behavior into consideration and improved diving strategy. The results show that the theory is still applicable in social networks. Moreover, messages are divided into different topics. Video and movie are found to have the highest applicability of the Spiral of Silence theory, while society topic has weak applicability.

Based on the theory, we try to predict polarity of users' opinions without considering semantic meaning in social networks. Rich features are extracted, and the results indicate that opinions of users are related to environment. Sentiment of the parent node is found to have a significant influence on polarity of its child nodes.

And we also extend the theory of Spiral of Silence and demonstrate that there exists the Anti-Spiral of Silence effect in social networks from two real cases. Anti-Spiral of Silence is found to have positive effects in leading rational thinking and revealing truth of events.

A large amount of cascades reflecting the propagation trajectory of hot events are extracted to analysis effects of the Spiral of Silence. However, an event may include many cascades and we will analyze the theory through many cascades related to an event in the future.

Acknowledgement This work was supported by the Science and Technology fund of Guangdong Province (No. 2015B010131003), Major and Special Project of Collaborative Innovation on the Integration of Industry, Education and Research; Guangzhou (No. 201604010017) Collaborative innovation major projects. The authors thank the reviewers for their time to help them.

References

1. The Statistics Portal: Most famous social network sites worldwide as of September 2016, ranked by number of active users (in millions)
2. Noelle-Neumann, E.: Return to the concept of powerful mass media. Stud. Broadcast. **12**, 67–112 (1973)
3. Scheufle, D.A., Moy, P.: Twenty-five years of the spiral of silence: a conceptual review and empirical outlook. Int. J. Public Opin. Res. **12**(1), 3–28 (2000)
4. Fashuo Wang: Process of citizen participation in public policy by network. PhD thesis, Fudan University (2012)
5. Luo, L., Li, M., Wang, Q., et al.: Spiral of silence in social networks: a data-driven approach. In: IEEE/ACM International Conference on Advances in Social Networks Analysis and Mining (2016)
6. Long, J.-Y., Zhou, K.: Revelation of silent spiral theory and anti-silent spiral for spread of socialist core value system. J. Huazhong Univ. Sci. Technol. (Soc. Sci. Ed.) **2013**, 13–19 (2013)
7. Moy, P., Domke, D., Stamm, K.: The spiral of silence and public opinion on affirmative action. J. Mass Commun. Q. **78**(1), 7–25 (2001)
8. Crandall, H.M., Ayres, J.: Communication apprehension and the spiral of silence. J Northwest Commun. Assoc. **31**, 27–39 (2002)
9. Gearhart, S., Zhang, W.: Gay bullying and online opinion expression testing spiral of silence in the social media environment. Soc. Sci. Comput. Rev. **32**(1), 18–36 (2014)
10. Matthes, J., Morrison, K.R., Schemer, C.: A spiral of silence for some: attitude certainty and the expression of political minority opinions. Commun. Res. **37**(6), 774–800 (2010)

11. Zuercher, R.: In my humble opinion: the spiral of silence in computer-mediated communication. In: Proc. of the 94th Annual National Communication Association (NCA) Convention (2008)
12. Lee, N.Y., Kim, Y.: The spiral of silence and journalists' outspokenness on Twitter. Asian J. Commun. **24**(3), 262–278 (2014)
13. Wu, Y., et al.: Exploring the spiral of silence in adjustable social networks. Int. J. Mod. Phys. **26**, 1550125 (2015)
14. Thomas, Z., Fawzi, N.: Can online exemplars trigger a spiral of silence? Examining the effects of exemplar opinions on perceptions of public opinion and speaking out. New Media Soc. **19**(7), 1034–1051 (2016)
15. Hampton, K.N., Rainie, L., Lu, W., et al.: Social Media and the 'Spiral of Silence'. Pew Research Center, Washington, DC (2014)
16. Tan, C., Lee, L., Tang, J., Jiang, L., Zhou, M., Li, P.: User-level sentiment analysis incorporating social networks. In: KDD'11, pp. 1397–1405 (2011)
17. Pang, N., Ho, S.S., Zhang, A.M.R., et al.: Can spiral of silence and civility predict click speech on Facebook? Comput. Hum. Behav. **64**, 898–905 (2016)
18. Guo-hua, W., Yu-lu, D.: Study on the phenomenon of "anti- silence spiral" in network communication. J. Beijing Inst. Technol. (Soc. Sci. Ed.) **12**(6), 116–120 (2010)
19. Alias-i: LingPipe 4.1.0. http://alias-i.com/lingpipe (2016)
20. Bastian, M., Sebastien, H., Mathieu, J.: Gephi: an open source software for exploring and manipulating networks. In: Proc. of 3rd International AAAI Conference on Weblogs and Social Media (ICWSM) (2009)
21. Hu, Y.: Efficient, high-quality force-directed graph drawing. Math. J. **10**(1), 37–71 (2005)
22. Newman, M.E.J., Girvan, M.: Finding and evaluating community structure in networks. Phys. Rev. E. **69**(2), 026113 (2004)
23. Cheng, J., Adamic, L.A., Dow, P.A., Kleinberg, J., Leskovec, J.: Can cascades be predicted? In: Proceedings of the 23rd International Conference on World Wide Web, pp. 925–936. ACM (2014)
24. Blondel, V.D., Guillaume, J.-L., Lambiotte, R., Lefebvre, R.: Fast unfolding of communities in large networks. J. Stat. Mech. Theory Exp. **2008**, P10008 (2008)
25. Kubat, M., Matwin, S.: Addressing the curse of imbalanced training sets: one-sided selection. In: Proceedings of the Fourteenth International Conference on Machine Learning, vol. 97, pp. 179–186 (1997)

Prepaid or Postpaid? That Is the Question: Novel Methods of Subscription Type Prediction in Mobile Phone Services

Yongjun Liao, Wei Du, Márton Karsai, Carlos Sarraute, Martin Minnoni, and Eric Fleury

Abstract In this paper, we investigate the behavioural differences between mobile phone customers with prepaid and postpaid subscriptions. Our study reveals that (a) postpaid customers are more active in terms of service usage and (b) there are strong structural correlations in the mobile phone call network as connections between customers of the same subscription type are much more frequent than those between customers of different subscription types. Based on these observations, we provide methods to detect the subscription type of customers by using information about their personal call statistics, and also their egocentric networks simultaneously. The key of our first approach is to cast this classification problem as a problem of graph labelling, which can be solved by max-flow min-cut algorithms. Our experiments show that, by using both user attributes and relationships, the proposed graph labelling approach is able to achieve a classification accuracy of ~87%, which outperforms by ~7% supervised learning methods using only user attributes. In our second problem, we aim to infer the subscription type of customers of external operators. We propose via approximate methods to solve this problem by using node attributes, and a two-way indirect inference method based on observed homophilic structural correlations. Our results have straightforward applications in behavioural prediction and personal marketing.

Keywords Subscription types · Mobile phone services · Behavioural prediction

Y. Liao · M. Karsai (✉) · E. Fleury
Univ Lyon, ENS de Lyon, Inria, CNRS, UCB Lyon 1, LIP UMR 5668, IXXI, Lyon, France
e-mail: yongjun.liao@inria.fr; marton.karsai@inria.fr; eric.fleury@inria.fr

W. Du
Univ Lyon, INSA Lyon, Inria, CITI, Villeurbanne, France
e-mail: wei.du@insa-lyon.fr

C. Sarraute · M. Minnoni
Grandata Labs, San Francisco, USA
e-mail: charles@grandata.com; martin@grandata.com

© Springer International Publishing AG, part of Springer Nature 2018
M. Kaya et al. (eds.), *Social Network Based Big Data Analysis and Applications*,
Lecture Notes in Social Networks, https://doi.org/10.1007/978-3-319-78196-9_8

1   Introduction

In most of the countries, mobile phone operators propose two subscription options for their customers. In case of *prepaid* subscription, credit is purchased in advance by the customer and access to the service is granted only if there is available credit. On the contrary, in case of *postpaid* subscription, a user is engaged in a long-term contract with the operator, and service is billed according to the usage at the end of each month. Typically, a contract specifies a limit or "allowance" of minutes and text messages for what a user is billed at a flat rate, while any further usage incurs extra charges. Due to these differences in the level and time of engagement, different options of subscriptions may be adopted by typical user groups characterised by similar age, location, socioeconomic status, etc. Moreover, due to effects of homophily and social influence, these people may be even connected to each other and communicate frequently, thus forming locally homogeneous sub-structures in the larger mobile communication network. Such correlations between the communication dynamics, structure, and customer features can potentially be used to differentiate between customers.

In this paper, we are interested in such behavioural differences of prepaid and postpaid customers for at least three reasons.

- The detection of typical patterns of service usage corresponding to different subscription types may provide useful information for the operator when planning network management and service pricing.
- The identification of customers with atypical service usage patterns of their actual subscription type may help the design of better direct advertising and personal services.
- The inference of the subscription type of customers of other operators could help direct marketing to convince customers to migrate to the actual provider.

Our study based on the analysis of customers of a single operator reveals that the main difference between prepaid and postpaid users is in the way they use mobile phone services. As a first result, we found that postpaid users are more active and on average make more calls, toward more people as compared to prepaid users. We also found that only relying on attributes of call statistics, such as total duration and number of outgoing calls, we can classify fairly well customers, with an overall accuracy of ∼80%, by using standard machine learning tools.

However, to achieve better results, further we investigated the proxy social network of users mapped out from their mobile phone communication events. This analysis revealed strong structural correlations among users of the same subscription type, who actually called each other ∼3 times more frequently than others holding the opposite subscription type. This observation suggested that the classification accuracy could potentially be improved by algorithms considering not only individual activity patterns but also relationships between users at the same time. This is important, as, on the one hand, classification methods typically focus on finding distinct attributes associated to each class while ignoring the network topology; and on the other hand, methods aiming to partition networks (e.g. tools of community detection) typically ignore attributes associated to each node. By considering simultaneously the two sources of information, we propose an approach to combine classification and graph partitioning techniques in a unified framework. To this end, we formulate the classification problem as a problem of graph labelling, that is assigning a label, either prepaid or postpaid, to each user by taking into account both the user attributes and the structure in the mobile phone call network. This graph labelling problem can be solved by algorithms such as max-flow min-cut [4, 14], which in turn lead to an improved classification accuracy of 87%.

Inspired by these results, we further addressed the practical problem of inferring the service type of customers of external operators from the view of the host provider. Such information is important for mobile service operators to expand business in the already crowded telecommunication market. In these settings, we consider two sets of customers: one set includes customers of the host operator (*company users*) with the subscription type and all call records available. The other set includes customers of external operators (*non-company users*) with no information available but only about their interactions with company users. As mobile service operators never share the data of their customers for the concerns of economy, security and privacy, an operator can only perform this inference problem by using visible data such as the subscription types and the communication records of its own company users (including calls with non-company users) but without knowing anything about the communications between any two non-company users. However, by assuming that the same structural correlations exist between customers of different providers as observed between company users, we propose possible ways to infer and indirectly verify subscription types of non-company users, even without knowing their real values. This explanatory study below shows that our direct inference method, based on node attributes and structure extracted from call data between two operators, alone can achieve an accuracy of ~70%, while indirect inference methods may also provide meaningful insights.

The rest of the paper is organised as follows. In Sect. 2, we discuss the related works, in Sects. 3 and 4 we describe the dataset in use and the classification results using machine learning methods based merely on node attributes. Section 5 presents our graph labelling algorithm which exploits both node attributes and the local network structure. In Sect. 6 we introduce our methods of inference of the subscription types of non-company users. Finally in Sect. 7 we conclude our work and discuss some potential future directions of research.

2 Related Work

Our work is motivated by the work of Yang et al. [19] in which network communities are identified by using both user attributes and the structure of the social network. However, one important distinction of the present work is that in [19] the authors aimed to cluster nodes with similar attributes into some apriori unknown number of communities, whereas we aim to find discriminating attributes of nodes and assign them to two pre-defined communities. In addition, our approach is based on graph labelling and solved by simple graph algorithms, whereas [19] is based on a generative model optimised by block-coordinate descent. Other related works on community detection with user attributes include [5, 15, 17]. However, all of them focus on clustering nodes in a network into priori unknown partitions or communities, while none of them address the classification of nodes into pre-defined classes as here. A related classification problem which has been well studied in the literature is the inference of user demographics such as nationality, gender and age from social and mobile call networks [2, 6, 8, 12, 18]. While interesting, most of the previous work exploited only node attributes, some of which are computed from local connections.

Knowledge transfer from one network to another is also a hot topic in the field of machine learning and social network analysis. Most of these works focused on the inference of the structure, that is link prediction, using information from multiple networks. For example, Eagle et al. inferred the friendship network from mobile call data [9] and Dong et al. [7] and Ahmad et al. [1] predicted links across heterogeneous networks, which may only be partially visible. In addition, Tang et al. [16] studied the problems of relationship classification across networks and Kong et al. [11] inferred the anchor links between users in different social networks, that is the same user with different accounts. In contrast to most previous work, our study on cross-network inference focuses on the classification of nodes in the targeted network using only the connections between two networks.

3   Dataset

Our dataset is a sequence of anonymised Call Detailed Records (CDR) collected by a single operator in a Latin American country. It contains ~280 millions of call and SMS events recorded during 1 month between ~20 millions of mobile phone users. Each CDR records the starting time of the event (date and time), communication type (call or SMS), the call duration (in case of call event), the anonymised identifier of the user originating the call (caller), and the anonymised identifier of the person receiving the call (callee), but not the content of that transaction. Note that from the caller and callee at least one is a customer of the operator.

For the first part of our study, we constructed a dataset (DS1) of 3.2 million users, who are actual customers of the operator, and have available information about their subscription type being prepaid or postpaid. After filtering out users with either too few (total duration of calls $<10\,\mathrm{s}$) or unrealistically many (total duration of calls $>100,000\,\mathrm{s}$) communications, we obtained about 1.3 million prepaid and 1.2 million postpaid users in total. For the purpose of our investigation, from the 190 million CDRs of 2.5 million selected users, we constructed a directed weighted communication network with 90 million links between interacting users. More precisely, the communication network $G = (V, E)$ was defined by mobile users as nodes connected by directed weighted links. Links $(u, v, w) \in E$ between users u and v (where $u, v \in V$) were drawn if at least one communication event took place between them during the observation period. The weight w of a directed link was defined as the number of communication events or the total duration of calls that took place between u and v (actual weight definition is specified later). Note that DS1 is balanced with roughly the same number of users for each type. However, our approach does not rely on this fact as the only assumption we take is based on a structural correlation in the communication network, which we will verify in the next sections.

For the second part of our study, we considered the whole available communication sequence to construct a dataset (DS2). After similar filtering as before, we obtained 3.1 million company users (1.6 million prepaid and 1.5 million postpaid), with full list of CDRs available, and 14.1 million non-company users, with CDRs available only with company users. Note that the increase of company users in DS2 as compared to DS1 is due to the company users who only interacted with non-company users, thus appear only in DS2 with non-zero degree. In DS2, in total we had 190 million calls between company users and 90 million calls between company and non-company users. Using these events, we constructed a network, which contained 90 million links between company users (just as in DS1), and 44.6 million links between company and non-company users.

4   User Attributes and Classification

For each user, using the first dataset DS1 we extracted a number of attributes related to call statistics including

- Total number of outgoing calls
- Total duration of outgoing calls
- Mean and standard derivation of the duration of outgoing calls
- Outgoing degree (number of callees) in the CDR network

These attributes turned out to be very different for prepaid and for postpaid users as shown in Fig. 1 where the distributions of (a) the total number of outgoing calls, (b) the total and (c) the average duration of outgoing calls and (d) the outgoing degrees are depicted. From these measures, we found that on average postpaid users make 2.9 times more calls to 2.5 times more people as compared to prepaid users. Note that we tested and found that attributes related to incoming calls and SMS are not informative. By using these attributes, we built our classifiers with different machine learning methods including support vector machines (SVM), Boosting (AdaBoost), Naive Bayes and Decision Tree to estimate the subscription type of each user. In every experiment presented in the paper, the classifiers were trained using 20,000 randomly sampled users, half of them prepaid and half postpaid, and tested on the rest of the 2.5 million users. After these experiments, we found that SVM and AdaBoost achieved the best accuracy of about 80%, while Naive Bayes achieved

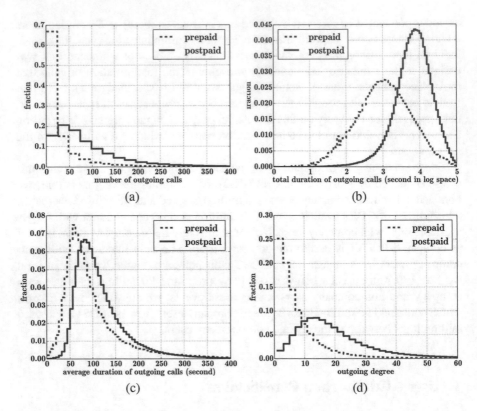

Fig. 1 Distribution of user call activity attributes as (**a**) total number of outgoing calls, (**b**) the total (in the logarithmic space) and (**c**) average duration of each outgoing call and (**d**) the outgoing degree of each user

Table 1 Confusion matrix by SVM

Accuracy = 0.803	Postpaid	Prepaid
Postpaid	0.822	0.178
Prepaid	0.214	0.786

about 77% and Decision Tree about 71%. The confusion matrix (each column of the matrix represents the instances in a predicted class while each row represents the instances in an actual class) for SVM is given in Table 1.

Note that the Naive Bayes method is a probabilistic learning method providing a probability for each node to belong to one of the subscription type, unlike SVM and AdaBoost, which provide binary values for labelling. For this reason, although less accurate, Naive Bayes is used later to define our graph labelling approach.

Besides the above mentioned attributes, which are related only to call statistics, attributes extracted from the social-communication network may also help our

Table 2 Proportion of calls within and between the two types of users

Callee \ Caller	Prepaid	Postpaid
Prepaid	0.791	0.209
Postpaid	0.173	0.827

Table 3 Confusion matrix for SVM with the proportion attributes

Accuracy = 0.891	Postpaid	Prepaid
Postpaid	0.897	0.103
Prepaid	0.115	0.885

prediction. The analysis of the network revealed a strong homophilic structural correlation, as calls between the same type of users (prepaid or postpaid) appeared on average three times more often than the ones across the two types of users. The matrix disclosing the fraction of calls by the two types of users is given in Table 2. This highly partitioned structure suggests that potentially the classification accuracy can be further improved by exploiting the observed sparse connectivity between the two user sets.

To test our conjecture, we extracted postpaid portion attributes, by using the known subscription types of neighbouring users for each user. The portion attributes in focus were

$$ F_n^i = n_i^{po}/n_i, \qquad F_c^i = c_i^{po}/c_i, \qquad \text{and} \qquad F_d^i = d_i^{po}/d_i, \qquad (1) $$

where F_n^i (resp. F_c^i, and F_d^i) assigns the portion of the n_i^{po} number of postpaid users in the callee set (resp. the c_i^{po} number of calls, and the d_i^{po} duration of calls to postpaid users) and the n_i number of users in the callee set (resp. the c_i number of calls, and the d_i total duration of calls) of a user i. With these extra portion attributes we could largely improve the accuracy of SVM and AdaBoost from ~80% to ~89%. The confusion matrix for SVM is given in Table 3.

However, in real settings it is impossible to measure these portion attributes, as they require the knowledge of the service type of each user, which are to be estimated. This way the outcomes provided by these methods are only informal in terms of our original classification problem as they use information what are assumingly not available. On the other hand, these methods can be used to detect false positive cases, where a user was assigned with a subscription type, while having another type of contract. This information can be used for direct marketing to provide services to costumers, which might better fit their needs. In any case, the positive results of this initial analysis motivated us to exploit the network topology in our new classification method defined in the next section.

5   Classification with User Attributes and Network Topology

5.1   Graph Labeling

The key idea here is to exploit both user attributes and the network structure to detect the service type of each user. This is to be done without any prior knowledge about the subscription types of users. To do so, we cast the classification problem as a problem of graph labelling and show that this graph labelling problem is equivalent to the classic network flow problem, namely max-flow min-cut. According to the max-flow min-cut theorem [14], in a flow network, the maximum amount of flow passing from the source s to the sink t is equal to the minimum cut, that is the smallest total cost of the links, which if removed would disconnect the source s from the sink t.

To see the connection between the two seemingly irrelevant problems, we take the CDR network $G = (V, E)$ and construct a new graph $G' = (V', E')$. We add two auxiliary nodes $V' = V \cup \{s, t\}$: these two nodes represent the two labels of prepaid and postpaid, which are assigned as the source s and the sink t, respectively, as illustrated in Fig. 2. Initially, we connect each user to both label nodes: $E' = E \cup \{(s, v) \mid \forall v \in V\} \cup \{(u, t) \mid \forall u \in V\}$. We call the links $(e = (u, v), u, v \in V)$ between users as *social links* and the links $(e = (s, u)$ or $e = (u, t), u \in V)$ between any user and one of the two label nodes as the *labelling links*. The costs associated to the social and labelling links will be described in the next section. However, independent of the definition of the cost functions, the following theorem guarantees that the minimum cut of the graph in Fig. 2 assigns a label to each user.

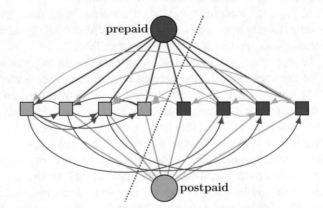

Fig. 2 Graph labelling for the classification of the service type of each user. Square nodes represent users and circle nodes represent the labels of prepaid and postpaid. Each user is connected in the mobile call network and with both label nodes. A feasible cut will partition the graph so that no flow can go from the source (prepaid) to the sink (postpaid). The minimum cut (the dashed line) is the feasible cut with the least overall cost

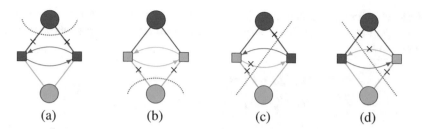

Fig. 3 Four possible cases for the feasible cut of the subgraph of a pair of connected users. Note that the labelling link which was cut indicates the labelling of a user. In the last two cases, only the one social link with a cross sign is cut. (**a**) Case 1. (**b**) Case 2. (**c**) Case 3. (**d**) Case 4

Theorem 1 *The minimum cut of a graph like in Fig. 2 will cut one and only one labelling link of the two labelling links of each user.*

Proof By definition, a cut is feasible if it disconnects the source and the sink, and the minimum cut is the feasible cut with the least total cost. It is trivial that any feasible cut will cut at least one of the two labelling links of each user; however, we need to show that any feasible cut that cuts both labelling links from a user is not a minimum cut of the graph. To see this, we consider a subgraph with a pair of doubly connected users. It is clear that the minimum cut of this subgraph can only be one of the four possible cases as shown in Fig. 3, all of which cut only one labelling link of each node. Which of these cuts apply in an actual subgraph depends on the cost defined on each link, but in any case, any graph containing such subgraphs must have the minimum cut which holds the same condition.

It is important to note that the labelling link which is cut between a node and a label indicates the labelling of the node, which reflects the intuition that cut links tend to have smaller costs.

5.2 Cost Functions and Problem Formulation

In order to extend our classification problem as a graph labelling or max-flow min-cut problem, we need to associate a cost to each link in Fig. 2. To this end, we use the information about both user attributes and relationships encoded in the social structure. For user u, let x_u be the user attribute vector, extracted from call statistics as discussed in Sect. 4, and f_u be the possible label of either prepaid or postpaid, represented by 0 and 1, respectively. On the one hand, we use Naive Bayes to compute the probability $P(f_u|x_u)$ that user u is a prepaid ($f_u=0$) and postpaid ($f_u=1$) subscriber. Note that Naive Bayes was adopted as it provides the probability of the classification rather than just a binary classification result. Subsequently, we define the D_u cost of each labelling link based on the the probability function $P(f_u|x_u)$ as

$$D_u(f_u = 0) = -\log(P(f_u = 0|x_u)).$$
$$D_u(f_u = 1) = -\log(P(f_u = 1|x_u)). \tag{2}$$

This definition assigns a small cost to a labelling link if the probability of a user taking the corresponding label is large. Note that in Naive Bayes, the normalisation condition holds, that is $P(f_u = 0|x_u) + P(f_u = 1|x_u) = 1$, while we force $P(f_u|x_u) > 1e - 10$ to avoid zero values in the logarithm of Eq. (2).

On the other hand, to define the cost of a social link, we rely on our earlier observations that postpaid users tend to have 2.5 times larger k_{out} outgoing degrees as compared to prepaid customers, and that the two types of users are sparsely connected with each other. The outgoing degree $k_{\mathrm{out}}(u)$ of each user u is a simple structural attribute, which turned out to be discriminative here. Thus, we define the cost of a social link $w_{(u,v)}$ as a slightly modified Ising model [10], that is,

$$w_{(u,v)}(f_u = 0, f_v = 0) = w_{(u,v)}(f_u = 1, f_v = 1) = 0,$$
$$w_{(u,v)}(f_u = 0, f_v = 1) = \frac{1}{k_{\mathrm{out}}(v)},$$
$$w_{(u,v)}(f_u = 1, f_v = 0) = \frac{1}{k_{\mathrm{out}}(u)}. \tag{3}$$

This definition assigns zero cost to a link, which connects two users of the same label (corresponding to the first two cases in Fig. 3a, b). Otherwise, when the label is different, a small cost is assigned that is inversely proportional to the outgoing degree of the user labeled as a postpaid customer (corresponding to the last two cases in Fig. 3c, d). Note that we also tried to define the cost as a measure of social strength between connected users, such as the number of outgoing calls or duration of outgoing calls, but obtained worse performance.

With the above defined cost functions for each social link (u, v), the graph labelling problem in Fig. 2 can be interpreted as an energy minimisation problem formulated as

$$\min \sum_{u \in V} D_u(f_u) + \lambda \sum_{u,v \in V} w_{(u,v)}(f_u, f_v). \tag{4}$$

Interestingly, Eq. (4) can be associated to a class of discrete optimisation problems, which has been well studied in the field of computer vision and machine learning [4]. In this terminology, the first term D_u, called the data term, reflects the cost of the labelling link between u and label f_u, while the second term is the smoothness term, reflecting the costs of social links between users u and v, which encourages connected users to take the same labels. Note that λ is a parameter that controls the trade-off between the two terms and its impact on accuracy will be described in the next section. Also note that Eq. (4) can be solved efficiently by any standard max-flow min-cut methods among which the push-relabel algorithm was chosen. In addition, we remark that the cost functions defined in Eqs. (2) and (3) do not require the labels of the nodes as inputs as they merely rely on probabilities determined by the Naive Bayes method using user attributes other than portion attributes.

5.3 Implementation and Results

Based on the above defined model formulation, we can solve the graph labelling problem using standard max-flow min-cut algorithms [3]. Our experiments showed that the graph labelling approach achieved an accuracy of $\sim$84% which is better than using supervised learning on user attributes alone. Note that the parameter λ has an impact on the classification accuracy. If $\lambda = 0$, we ignore completely social relationships and graph labelling degenerates to Naive Bayes using user attributes, while if $\lambda = +\infty$, we ignore completely user attributes and enforce all nodes to take the same label, resulting in an accuracy of $\sim$50% due to the similar sizes of two types of users in our dataset. Thus, we tuned λ by cross validation to our dataset and found the best value to be 100. Enlarging λ by 10 times resulted in an accuracy of $\sim$74%, whereas shrinking λ by 10 times gave us $\sim$80%.

In addition, we tried to prune the graph by fixing the label of some users under two conditions: if the probability of a user belonging to a class was larger than a threshold τ_1 and if the average probability of its neighbours belonging to the same class was larger than another threshold τ_2. In our experiments, we set $\tau_1 = 85\%$ and $\tau_2 = 65\%$ which led to fixed labels for about 16% of users. This way of pruning improved the accuracy of our prediction to 87%. The confusion matrix of this method is given in Table 4, while Table 5 summarises the accuracy achieved by all different methods applied so far.

Table 4 Confusion matrix by our graph labelling approach

Accuracy = 0.868	Postpaid	Prepaid
Postpaid	0.901	0.099
Prepaid	0.163	0.837

Table 5 Classification accuracy of different methods using only user attributes

Method	Accuracy
Graph labeling with graph pruning	0.868
Graph labeling	0.844
SVM with user attributes	0.803
Naive Bayes with user attributes	0.770

Note that the runtime of our approach is about 0.5 h on a desktop with 32G memory and CPU Xeon of 1.90 GHz which is not that slow considering the size of our graph. In particular, we adopted the push-relabel method to solve the max-flow min-cut problem, which has the computational complexity of $O(V^2E)$, where V and E are the number of nodes and edges, respectively. Also note that, our graph labelling algorithm utilises the Naive Bayes for modelling the probability that a user is a prepaid or postpaid subscriber. This calculation requires 20,000 randomly sampled training data, as mentioned earlier. However, in the optimisation, we assign labels to all 2.5M users including ones in the training set as well. The training samples have no impact to the results as they are less than 1% of the total users and they are only used for learning the cost of labelling links.

6 Subscription Type Inference of Non-company Users

In most of the countries, typically there are several mobile service providers, who compete for customers on the same market. However a provider has access to user data (e.g. subscription type) of its own customers only, while information about the users of other providers are limited to the CDRs between the company and non-company customers, recorded for billing purposes. In this section, we use such CDRs to introduce two methods to infer the subscription type of non-company users from a company point of view. Such knowledge is certainly valuable for a provider, for example to design advertisement strategies to motivate the churning of non-company users.

However, this is a difficult problem as the direct verification of inferred subscription types of non-company users is not possible. To come around this problem, we build on the assumption that call feature patterns, which were shown to be relevant in our classification problem earlier (see Sect. 4), actually characterise all customers on the market, not only the company users. This way such features may provide predictive power when characterising calling patterns between company and non-company users, thus can be effectively used for inference and help the design of new validation methods.

6.1 Inference Using Call Statistics

In our first method, let's consider two mobile providers: A represents our central company with available information about its company users, while B is an external company (of non-company users) with no user information available. First we check whether characteristic call features, observed in Sect. 4, also differentiate between prepaid and postpaid users in A but considering only inter-company communication. Here we take each user in A from DS2, and using their CDRs only with users in B, we measure the distributions of the number and duration

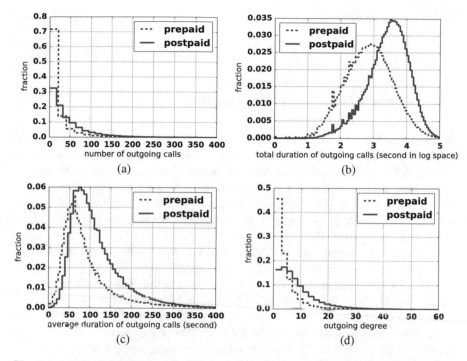

Fig. 4 From left to right are the distributions of each user for the total number of outgoing calls, the total (in the logarithmic space) and average duration of each outgoing call and the outgoing degree of each user. Note that these attributes are computed using the data of calls from the operator network of A to that of B, whereas those in Fig. 1 are computed using the data of calls within the operator network of A

of outgoing calls and their out-degree. As shown in Fig. 4, these distributions yet appear to be very different for prepaid and postpaid A users, just as it was shown earlier for intracompany communications. It suggests that these node attributes are not subjective to intracompany communications only but may generalise for any communication tie, even between users of different providers.

Based on these observations, we assume that these features characterise symmetrically the communications between any users independent of their provider. However, since we have access to the information of A users only, this is the only set we can use for direct verification of the results of an inference method. Thus in our first approach we took the point of the company B, and we actually try to infer the subscription types of users in A, by only using CDRs between the users of A and B and the characteristic attributes shown in Fig. 4. We applied the same machine learning methods with the same setting as described in Sect. 4, but now on the dataset DS2. These experiments showed that our method based merely on statistical node attributes alone can achieve an accuracy of $\sim$68.2% using SVM and Boosting. The confusion matrix by Boosting is given in Table 6. It demonstrates that using only inter-company communications and user attributes, one can infer the subscription types of customers of another provider with relatively high accuracy.

Table 6 Confusion matrix
by boosting

Accuracy = 0.687	Postpaid	Prepaid
Postpaid	0.682	0.318
Prepaid	0.317	0.693

6.2 Inference Using Label Propagation

In our earlier data analysis, in Sect. 4, we found that calls between the same type of users (prepaid or postpaid) appeared ~3 times more frequently than between users holding different subscriptions. Although this was observed between only users of A, we may assume similar homophilic correlations to be present between users of A and B. Based on this assumption here, we present a second method to infer subscription types of non-company users using label propagation together with an indirect validation method.

Let's consider a bipartite network $G = (V_A, V_B, E_{AB})$, where V_A is the set of all A company users, V_B is the set of all B non-company users, and E_{AB} contains all directed links between users in V_A and V_B. A directed link is defined between $a \in V_A$ and $b \in V_B$ (resp. between b and a) if a called b (resp. b called a) at least once during the observation. In this network, we know the subscription types of all users in V_A, while we would like to infer the subscriptions of V_B users. As edges in E_{AB} are real social ties inferred from communication records, they are not independent but show degree correlations, could be strong ties of communities, and may show homophilic correlations as we mentioned earlier. The knowledge of subscription types on side A, together with the strong effects of subscription homophily, and the structurally correlated bipartite network provide us a potential way to define a label propagation method using the majority rule to infer the subscription types of non-company users in V_B.

More precisely, we take all users from V_B and their incoming ties, and assign them a label (prepaid or postpaid) corresponding to the majority of the subscription type of their connected friends in V_A. In case of equality, we selected randomly a label by respecting the overall balance between pre- and postpaid users observed in A. However, even we can assign labels for all the users in V_B, we cannot validate the precision of our inference as we do not have access to the subscription types of non-company users. On the other hand, assuming all correlations to be symmetric between the two sets of users, we can use the inferred labels of V_B users to assign labels to users in V_A applying the same majority rule inference method. After these two ways of label inference, we can validate the precision of the inferred labels of users in V_A by comparing them to their real subscription types. The result will give us a lower bound for the inference error in V_B as erroneous inference propagates also during the process, and the entropy of inference cannot increase during the two ways of inference. As a result we found that on average our method identified the correct labels of 96.4% of users in V_A. Our measurements were averaged over 100 independent realisations with results summarised in the confusion matrix shown in Table 7.

Table 7 Confusion matrix of label inference on the empirical structure

Accuracy = 0.964	Postpaid	Prepaid
Postpaid	0.961	0.039
Prepaid	0.034	0.966

Table 8 Confusion matrix of label inference on the randomised structure

Accuracy = 0.562	Postpaid	Prepaid
Postpaid	0.539	0.461
Prepaid	0.415	0.585

This method could give trivial results if two conditions are satisfied at the same time: the average degree in the bipartite network is 1, and all links are bidirectional. In our case, although the 89% of links found to be bidirectional, the average degree is relatively high $\langle k \rangle \simeq 2.593$ suggesting that the inferred labels are not trivial. This can be further validated by eliminating structural correlations from the bipartite network and see whether we can recover the original results. If the accuracy of inference remains the same in the uncorrelated network, our results are trivial and structural correlations do not play a role here. On the other hand if the accuracy drops in the uncorrelated networks, our results are meaningful. To remove structural correlations, we used a configuration network model scheme [13], that is, we choose random pairs of links of the same direction and swapped them while disallowing double links. Repeating this step several times, we received a random bipartite structure where the in- and out-degrees of nodes did not change but any structural correlation vanished from the network. Repeating our two-way label inference method on 100 of such independently generated random bipartite structures, we found that the precision dropped to 56.2%, with a standard deviation of 5%, which suggests that our original findings were significant. For precise results, see the confusion matrix in Table 8.

7 Conclusions and Future Work

In this paper, we presented novel methods to detect prepaid and postpaid customers in mobile phone datasets by exploiting both user attributes and observed structural correlations in the communication network. First we addressed the problem with known inference methods relying on subscription type labels known for some users. We showed that in this case SVM provides us the best solution with 0.891 accuracy.

Second, as the main contribution of our work, we addressed a more challenging problem, to infer subscription labels without prior knowledge of subscription types, but only relying on user attributes and observed structural correlations. We cast the classification problem as a problem of graph labelling and solved it by max-flow min-cut algorithms. With this novel methodology, we achieved a classification accuracy of ~87% which is about ~7% better than supervised learning on user attributes alone.

Third, we provided two methods to infer subscription types of non-company users. We showed that using node attributes alone we can achieve already ~70% accuracy of inference. In addition, exploiting present homophilic structural correlations, we obtained good accuracy of inference by using a two-way label propagation method.

Our results have certain limitations. As we explained, the incomplete knowledge of subscription types of non-company users makes difficult the full verification of any inference problems. We provided one possible indirect solution here but this should be developed further using more complete datasets. We also generalised some observations on node features and structural correlations for any node and link in the communication network, which should be verified. In addition, the two-way inference method provides more accurate results if the average out-degree of non-company users is high. For some reasons, we found its value relatively small, which limits the precision of our inference results. We also note that the indirect inference method works well in case of competing providers with similar subscription portfolio and user base, thus where our observations on company users generalise the best.

In the future, there are several avenues to explore. First, one could extend our approach from binary classification to multi-class classification, which can be similarly formulated as a graph labelling problem but with more than two label nodes. Such problems are known to be NP-hard but greedy algorithms based on max-flow min-cut have been proposed to achieve approximate solutions [4]. Second, our dataset is intrinsically dynamical and contains CDRs over 1 month, what we all used to predict user subscription types. It would be interesting to study the impacts of the dynamics of the mobile call data over time for longer periods to understand how temporal patterns of communications can improve our predictions. Possible structural biases in the network between users of different customers could also be a direction to explore. Our aim here was to contribute to the more general discussion of user attribute inference and to give possible directions for potential future applications in behavioural prediction and personal marketing.

Acknowledgements We acknowledge Grandata to share the data and M. Fixman for his technical support. This research project was partially granted by the SticAmSud UCOOL project, INRIA, SoSweet (ANR-15-CE38-0011-01) and CODDDE (ANR-13-CORD-0017-01).

Authors' Contribution All authors read and verified the manuscript. YL, WD, MK, CS and EF participated in the writing; YL, WD, MK and EF designed the research; and YL run all numerical calculations.

References

1. Ahmad, M.A., Borbora, Z., Srivastava, J., Contractor, N.: Link prediction across multiple social networks. In: 2010 IEEE International Conference on Data Mining Workshops, Dec 2010
2. Bin, B., Milad Shokouhi, M.K.T.G.: Inferring the demographics of search users. In: 22nd International World Wide Web Conference (2013)

3. Boykov, Y., Kolmogorov, V.: An experimental comparison of min-cut/max-flow algorithms for energy minimization in vision. IEEE Trans. Pattern Anal. Mach. Intell. **26**(9), 1124–1137 (2004)
4. Boykov, Y., Veksler, O., Zabih, R.: Fast approximate energy minimization via graph cuts. IEEE Trans. Pattern Anal. Mach. Intell. **23**(11), 1222–1239 (2001)
5. Burton, S.H., Giraud-Carrier, C.G.: Discovering social circles in directed graphs. ACM Trans. Knowl. Discov. Data **8**(4), 21 (2014)
6. Chen, X., Wang, Y., Agichtein, E., Wang, F.: A comparative study of demographic attribute inference in twitter. In: Proceedings of the Ninth International Conference on Web and Social Media (2015)
7. Dong, Y., Tang, J., Wu, S., Tian, J., Chawla, N.V., Rao, J., Cao, H.: Link prediction and recommendation across heterogeneous social networks. In: Proceedings of the 2012 IEEE 12th International Conference on Data Mining (2012)
8. Dong, Y., Yang, Y., Tang, J., Yang, Y., Chawla, N.V.: Inferring user demographics and social strategies in mobile social networks. In: SIGKDD (2014)
9. Eagle, N., Pentland, A.S., Lazer, D.: Inferring friendship network structure by using mobile phone data. Proc. Natl. Acad. Sci. **106**(36), 15274–15278 (2009)
10. Kolmogorov, V., Zabih, R.: What energy functions can be minimized via graph cuts? IEEE Trans. Pattern Anal. Mach. Intell. **26**(2), 65–81 (2004)
11. Kong, X., Zhang, J., Yu, P.S.: Inferring anchor links across multiple heterogeneous social networks. In: Proceedings of the 22Nd ACM International Conference on Information and Knowledge Management (2013)
12. Malmi, E., Weber, I.: You are what apps you use: demographic prediction based on user's apps. CoRR (2016)
13. Newman, M.: Networks: An Introduction. Oxford University Press, Oxford (2010)
14. Papadimitriou, C.H., Steiglitz, K.: Combinatorial Optimization: Algorithms and Complexity. Prentice-Hall, Inc., Englewood Cliffs (1982)
15. Sun, Y., Aggarwal, C.C., Han, J.: Relation strength aware clustering of heterogeneous information networks with incomplete attributes. Proc. VLDB Endow. **5**(5), 394–405 (2012)
16. Tang, J., Lou, T., Kleinberg, J.: Inferring social ties across heterogenous networks. In: Proceedings of the Fifth ACM International Conference on Web Search and Data Mining (2012)
17. Xu, Z., Ke, Y., Wang, Y., Cheng, H., Cheng, J.: A model-based approach to attributed graph clustering. In: Proceedings of the 2012 ACM SIGMOD International Conference on Management of Data (2012)
18. Xu, H., Yang, Y., Wange, L., Liu, W.: Node classification in social network via a factor graph model. In: Advances in Knowledge Discovery and Data Mining. Springer, Berlin (2013)
19. Yang, J., McAuley, J., Leskovec, J.: Community detection in networks with node attributes. In: IEEE International Conference On Data Mining (ICDM) (2013)

Dynamic Pattern Detection for Big Data Stream Analytics

Konstantinos F. Xylogiannopoulos, Panagiotis Karampelas, and Reda Alhajj

Abstract The last two decades witnessed tremendous and astonishing developments in technology. This pushed for visible revolution in communication and electronics design leading to the production of computing devices of various sizes and capabilities, ranging from tiny sensors with limited specifications to mobile devices with huge power and rich functionalities, among others. These stimulated researchers and practitioners work hard seeking the best possible benefit from such novel devices to serve humanity. Gathering huge amounts of data is way easier and more affordable than ever before. Indeed, there is a clear shift from paper-based manual data collection to totally automated data collection even under sever conditions which were never feasible to consider before. Data is captured as a stream which may encapsulate some trends that may reveal certain aspects essential to our daily life. Identifying such trends in data streams is the main theme of the study described in this chapter. We mainly concentrate on real-time stream data analysis to better serve time-critical applications where instant decision making is crucial. This study builds on our methodology described in (Xylogiannopoulos et al. Frequent and non-frequent pattern detection in big data streams: an experimental simulation in 1 trillion data points. In: Advances in social networks analysis and mining (ASONAM), pp. 931–938, 2016) which considers detecting all repeated patterns in a big data stream. In the new dynamic approach, a sliding window is employed with LERP Reduced Suffix Array and the ARPaD algorithm to analyze one trillion digits composed from one million subsequences of one million digits each. We achieved like generating one data point every 300 ns.

Keywords Big data · Data stream · Pattern detection · LERP-RSA · ARPaD · Data analytics

K. F. Xylogiannopoulos (✉) · R. Alhajj
Department of Computer Science, University of Calgary, Calgary, AB, Canada

P. Karampelas
Department of Informatics and Computers, Hellenic Air Force Academy, Dekelia Air Base, Acharnes, Greece

© Springer International Publishing AG, part of Springer Nature 2018
M. Kaya et al. (eds.), *Social Network Based Big Data Analysis and Applications*,
Lecture Notes in Social Networks, https://doi.org/10.1007/978-3-319-78196-9_9

183

1 Introduction

Humans have realized the importance of data since their existence on earth. They invented various ways to keep and maintain data manually by various media and means. They continued to handle data manually until the early 1960s when computers were first used for storing and handling data electronically. Since then automated data capturing and maintenance improved gradually. This has been recently highly influenced by the rapid development in technology leading to the existence of a variety of communication devices capable of capturing data. These range from tiny sensors to powerful mobile phones, tablets, laptops, etc. Advanced communication technology has been well backed by emerging software communication platforms such as Facebook, Twitter, etc. Further, sensors have been installed in almost every device from washing machines to cars, among others. This revolution in hardware and software has brought up a situation where huge amounts of data could be captured electronically at high speed leading to data streams. Analyzing such streams could reveal various aspects related to the generating source and hence will help in studying and watching various devices properly by timely and appropriate decision making.

Data streams may encapsulate a variety of valuable trends which are highly important to understand the functionality, progress, performance, etc., of the particular devices generating the data. Capturing such trends is crucial and has motivated researchers to develop algorithms and data structures that could efficiently analyze data streams to find patterns of various characteristics. Some patterns might be outliers and may reveal some critical aspects or situations which if timely handled would save a device from malfunctioning and hence would lead to avoiding all associated consequences.

Data streams may be considered as a continuous time series which fast grow in size and hence require more effort from researchers to develop algorithms and structures capable of handling big data. Further, such streams may suffer from various problems which should be fixed or taken into consideration by the analyzing technique or algorithm. For instance, a stream may incorporate some noise or may be incomplete. Indeed, incompleteness cannot be avoided as data always flows and it is not possible to predict what will come next. The analysis of such series is not straightforward. It requires discretization first to create data structures more appropriate for further analysis. It is also possible to have the data analysis time independent, that is, it is irrelevant to produce the outcome based on certain time frame, and hence time complexity of the data analysis would not be considered critical. Examples of such analysis could be encountered in some popular domains like bioinformatics data, mathematics, among others.

For time-critical data analysis, it becomes crucial to have the best possible algorithm that utilizes a compact data structure for swift analysis to avoid any delay in the decision-making process. For instance, consider a stock market data which should be analyzed to predict the trend based on the discovered patterns. Investors will be interested in running an algorithm which allows them to act first.

Another example is weather forecast based on environment data analysis to predict possible unpleasant weather conditions so that people will be warned early before the disaster hits leading to causalities and severe damage. These applications and the like are time critical and hence require online and real-time analysis leading to frequent pattern analysis with high accuracy. The two most commonly used data structures in this domain are suffix trees and suffix arrays which represent all suffices of a string, that is, all substrings are enumerated iteratively by removing the head element each time to construct a new shorter suffix string. These two data structures have demonstrated significant advantages mainly in handling data which could be realized as a string. However, both data structures suffer from scalability problems when it comes to deal with and handle big data due to either static or dynamic data streams.

The work described in this chapter is intended to overcome the scalability problem and to contribute to the literature on big data stream analysis for pattern detection and analysis. For this purpose, we have used an innovative data structure, namely, Longest Expected Repeated Pattern Reduced Suffix Array (LERP-RSA) [1, 2] together with a novel algorithm which demonstrated extreme capabilities in analyzing big data streams, namely, All Repeated Patterns Detection (ARPaD) algorithm [1–3]. LERP-RSA is a data structure which has many significant advantages compared to other pattern detection-related data structures. It smoothly facilitates automatic classification and parallel execution. In other words, ARPaD is a novel algorithm that takes a string of arbitrary size as input and utilizes LERP-RSA data structure to detect all repeated patterns in the given string. The combination of these two pearls facilitates smoothly the fast analysis of big time series, or sequences in general, using a common personal computer.

The outcome from the aforementioned algorithm is a set of all possible patterns which can be further analyzed in real time and in parallel using minimum resources. The applicability, power, and effectiveness of the proposed methodology are demonstrated later in the experimental section by conducting two tests on a simulated big data stream constructed from the first trillion digits of π. In the first experiment, π will be divided into 1000 strings of length 1 billion, and each string will then be analyzed individually to detect repeated patterns. In the second experiment, the same dataset will be analyzed again using one million strings of length one million, but this time a dynamic sliding window technique will be used in order to have fast analysis directly on memory and to facilitate historical data analysis. It will not be possible to conduct such analysis by employing any of the existing methodologies described in the literature. This is true mainly due to the lack of data structures associated with existing methodologies to allow them to scale up for such a large dataset. Although some techniques may exist, yet, it was not possible to find any experimental results using such a big dataset for comparison purposes.

The rest of this chapter is organized as follows: Sect. 2 reviews the related work and describes the theoretical background that will be used for the proposed methodology. Section 3 describes the proposed methodology. Section 4 discusses the experimental analysis. Finally, Sect. 5 is conclusions and anticipated future work.

2 Related Work

2.1 Existing Methodologies

Two of the most commonly used data structures for pattern detections are suffix trees and suffix arrays. The suffix tree has been used for decades to deal with core string problems, for example, detecting all repetitions in a string [4], the longest repeated substring [5], etc. Moreover, several advanced algorithms have been introduced lately, for example, by Guo et al. [6] and Wu et al. [7], which address the wildcard pattern detection. The suffix array is an array data structure representing the positions of the lexicographically sorted suffix substrings of a string. It was introduced by Manber and Myers in 1990 together with a pattern detection algorithm of a specific substring existing in a string [8]. Based on that publication, several algorithms introduced for pattern detection which fostered the use of the original suffix array while some other researchers introduced several variations of the initial data structure with diverse advantages and disadvantages. Such an algorithm introduced by Franek, Smyth, and Tang for the detection of repeats used suffix arrays which satisfy specific initial conditions [9]. A first extending experimental application of the specific algorithm executed for the first time in 2008 [10] where Puglishi, Smyth, and Yusufu analyzed strings up to just 68 million characters. These methodologies rely heavily on the advantage of the suffix array data structure compared to the suffix tree, and, more specifically, the required space capacity which can be significantly smaller [8, 9].

Regarding the online data stream analysis, which is our focus in this chapter, some algorithms do exist which use different approaches and data structures than suffix arrays and suffix trees. Cormode and Hadjieleftheriou [11] have classified these algorithms in three categories: (1) Counter-based algorithms which monitor the new items coming as input and they decide whether the items will be kept and if yes to which counter they will be associated with, (2) Quantile algorithms which try to find the best allocation of quantiles and thus the corresponding frequency of the items, and (3) Sketch algorithms that use special data structures to linearly project the stream with the aim to solve the frequency estimation problem and consequently they use additional information to find the frequent items [11]. The counter-based category includes algorithms such as the Majority Algorithm of Boyer and Moore [12], Frequent Algorithm by Demaine et al. [13] and Karp et al. [14], Loosy counting by Manku and Motwani [15], and Space Saving by Metwally et al. [16]. The Quantile category incorporates algorithms such as the GK Algorithm of Greenwald and Khanna [17] and QDigest of Shrivastava et al. [18]. The Sketch algorithms category contains the CountSketch of Alon et al. [19] and CountMin Sketch of Cormode and Muthukrishnan [20].

All the aforementioned algorithms and many others that exist can perform pattern detection or pattern matching by having some kind of input arguments, that is, what kind of pattern, in most of the cases a string, we want to detect. The real problem that arises in pattern detection in general and data streaming analytics more particularly is to be able to detect patterns in an agnostic way, that is, without having any input argument or knowledge of what needs to be discovered. This is very important in several real-world applications, for example, in network security and data analytics, which also belong to the category of data streaming. In such a case, we do not know what we want to find in advance but instead we care to detect patterns as they occur. To the best of our knowledge, there is no algorithm in literature which allows pattern detection under these agnostic conditions except the All Repeated Patterns Detection algorithm (ARPaD) [1–3]. The detection of all repeated patterns could be proved extremely important because in combination with periodicity detection algorithms allow us to detect periodically repeated patterns, which are the fundamental target in forecasting, for example, in financial or weather time series. Additionally, the ARPaD algorithm, in contrast to association rules mining algorithms, does not use any frequency (support) thresholder and, in this way, can detect all repeated patterns by transforming the problem of frequent items detection to the more generic and almost impossible to solve for big data sets problem of detecting itemsets of any frequency. This can be done easily by using ARPaD to detect every pattern that occurs at least twice in just one run over the dataset and then the problem of finding patterns of any frequency can be transformed to a meta-analysis problem of the discovered patterns. This kind of data analysis is very efficient because it allows us to perform several meta-analyses without significant cost, compared to the pattern detection process, and without the need to re-execute the resource and time-consuming initial pattern detection process. Once all repeated patterns have been detected, then there is no reason to repeat the process while many other methodologies, for example, a priori and other association rules mining algorithms, require a new execution every time the initial parameters of the problem have changed. Predetermined models can be directly applied on the results to extract valuable information according to the needs of the analyst and the type of the dataset, while the meta-analyses of all repeated patterns that have been detected can be executed in parallel over the same resulting dataset producing the final outcome very quickly.

2.2 Theoretical Background

In order to analyze big data streams, the appropriate data structure needs to be utilized in order to minimize the memory requirements and the time of analysis. For that purpose, the LERP-RSA data structure will be used. The specific data structure has been proposed by the authors in [1–3] where the Longest Expected Repeated Pattern (LERP) has been defined. According to that, the LERP is defined as the longest pattern that is expected to be found at least twice in a string, or stream in our case. According to the Probabilistic Existence of Longest Expected Repeated Pattern Theorem which states that the probability of a string to be very long and occur twice in another string is extremely small [1, 2] and the LERP Calculation Lemma, an estimate of the value of LERP can be calculated using only the length of the string and the number of letters in the alphabet used [2, 11].

Another essential benefit by the utilization of the LERP-RSA is that it supports the classification of the data structure based on the alphabet used to build the string. As it has been presented in [1, 2], the classification level is defined as the exponent in which the alphabet should be raised in order to get the desired number of classes for the analysis. The specific characteristic of the data structure enables the proposed methodology to handle very large time series, by distributing the different classes in a cloud computing environment with multiple devices and as a result to allow the parallel execution of the algorithm that will be used for the analysis on each class. The following example demonstrates how the classification levels are used.

Let's suppose that the decimal system has been used to discretize a sequence of characters, then ten different classes from "0" up to "9" can be produced that start with the specific letter of the alphabet and include all the suffix strings starting with the specific character. Accordingly, the classification level 2 produces $10^2 = 100$ classes from "00" up to "99" and so on. The specific approach allows to significantly reduce the size of each class of the data structure since, for example, if we assume that the suffix strings are equally distributed in the string, then each class will be the 1/100 of the size of the original string. Consequently, for a string with length one billion characters, 100 classes of ten million characters each one can be created. Then the analysis can be carried out in all classes in parallel, provided that the appropriate hardware is available. Otherwise, the analysis algorithm can be executed in a semi-parallel mode, that is, each class can be analyzed sequentially and all repeated patterns can be detected. If the hardware configuration allows, that is, the number of CPUs, available RAM, disk space, etc., the process can be further optimized [2, 3, 21].

As briefly mentioned above, the ARPaD algorithm [3] will be used for the analysis and pattern detection. The specific algorithm not only is unique but also can perform better than other algorithms of pattern detection since (1) its time complexity is $O(n \log n)$ in the worst case $O(n)$ in the average case; (2) it is flexible to operate with LERP-RSA suffix arrays either stored on external media and database management systems, or on memory; (3) it can detect independently of the length, the string, or the alphabet all the repeated patterns found in a string; (4) by being able to use external media storage, it is possible to analyze the classes either in a different computing system or in different time spans based on the availability of the computing resources; (5) the size of the suffix array under analysis or of parts of the suffix array can be significantly reduced by applying clustering and classification techniques; (6) depending on the hardware availability, that is, processors, parallel implementation can be used by splitting the workload of the analysis to as many processors as possible; and (7) larger computing configurations such as network/cloud computing can be utilized to allocate the workload into several computers in a local network or in a cluster computing infrastructure [1, 2]. As a result and based on the aforementioned characteristics of the algorithm and the data structure, their combination can drastically reduce the overall time needed for the analysis and pattern detection, even for real-time analysis of big data streams.

Variations of LERP-RSA and the ARPaD algorithm that have been used in various other cases with significant results are for example in the detection of a

Distributed Denial of Service (DDoS) attack on a network server [22] detecting all repeated patterns in all the different parts of the IP addresses, or in the detection of frequent and non-frequent sequential itemsets [21], or for clickstream analysis [2], and in financial time series analysis [23]. The common characteristic of the aforementioned applications is that they can be transformed to data streams analytics problems. This transformation happens as continuous data points are entering the system dynamically. In [1, 2], it has been proved that in most of the above cases the produced sequences or strings are relatively small and thus they can be easily analyzed. The only case that is considerably different is the case of the DDoS attack, where in the effort to early detect and prevent an attack in the network, the analysis should be performed in real time. As it was proven in [22], it is possible to early detect and potentially prevent a DDoS attack using a combination of LERP-RSA and a variant of the ARPaD algorithm which achieved to complete the analysis faster than the actual attack that is, the escalation of the attack. The experiment carried out showed that a DDoS attack dataset of 150 million IP addresses simulating an attack of over 30 min was analyzed and in <4 min using a simple laptop computer.

In this chapter, a variation of the previously shown data streaming analytics approach [24] will be presented using a sliding windows technique. A synthetic dataset will be used to simulate a big data stream, which will be constructed by using one thousand strings of length one billion characters each one, derived from the first one trillion digits of number π. The digits of π can easily be replicated by any researcher who would wish to compare a new pattern detection technique in the future with the proposed one.

3 Proposed Methodology

The methodology proposed in this chapter attempts to address the problem of big data streaming analysis. Since it is very difficult to find big data streaming datasets freely available, we simulate the generation of streaming data by using one thousand different strings that are analyzed in real time. Each string consists of one billion digits and as it is proved later in the experimental analysis section, the real-time analysis of the specific dataset is equivalent to the analysis of data points generated every 2 μs. As it can be understood, the composite dataset used in the proposed methodology corresponds to faster data points production than weather or financial time series. Similarly, the proposed method can be used for multivariate analysis, that is, analyzing 1000 variables having one million data points each. This case, as it is again shown in the experimental analysis section, is equivalent with the production of a data point per 2 ms which also simulates a very fast generated time series. The following subsections describe two variations of the proposed approach. The first one has already been presented in [24], and it has been included in this chapter for completeness and comparison purposes with the latest one which is its advanced expansion.

Fig. 1 Methodology
execution diagram [24]

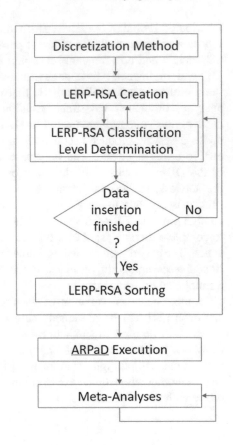

3.1 Sequential Execution

The first approach, as described in [24], is divided into three phases (Fig. 1). First, the data structure needs to be created. The specific construction requires three steps. The first step depends on the type of data points. If the data points are continuous values, then we need to decide how they will be discretized and what will be the length of each subsequence that will be used for our data entries. Let's suppose that we are interested in analyzing the stock prices and their data points are collected per second. Then, depending on our interest we may decide to analyze strings per minute (60 data points) or 10 min (600 data points) or in any other convenient length. Alternatively, for the benefit of the analysis, we may use variable size for each subsequence in order to conduct simultaneous analysis on different lengths. As it has been described in the theoretical background, the method requires defining the Classification Level to be used for the analysis based on the alphabet used for the discretization process. Then, the LERP value needs to be calculated for the construction of the LERP-RSA data structure which can be done easily using the LERP Calculation Lemma [1, 2] since we only need the length of each substring and the alphabet used for the discretization.

The first phase continues with the next step which involves the collection of the substrings of the LERP-RSA data structure. When a new data point is generated, then it is added in the reserved memory. If the memory already contains other data points, the new one is appended at the end of the list provided that the length of the string is less than the value of LERP. When the string stored in memory equals the value of the predefined LERP, the whole string is stored in the appropriate class based on the first digit(s) according to the classification level and then the digit that was inserted first is removed. At the specific stage of the process execution, we have stored in the system a substring with length equal to LERP and in the reserved memory a substring with length LERP—1 which waits for the next data point to enter the system. When the next data point is generated the same process is repeated, namely, the new string with length LERP is stored in the appropriate class and its first digit is removed from the reserved memory. The specific process will be repeated as many times as the length of the subsequence has been determined at the beginning of the process. At the end of the specific step, all the classes will have to be generated with as many substrings as the length of the subsequence as it has been determined in the first step.

Figure 2 illustrates the above mentioned process when seven digits of π have been generated starting at position 8 using as LERP the value 5. When all the first five digits have been generated, a substring of length 5 has been created. When the next digit is generated, in this case, digit 7, then the substring "53589" is stored in the class labeled "5," and then digit 5 is removed from the first position of the reserved memory and digit 7 enters at the end of the substring forming a new substring "35897." As mentioned above, the process is repeated for the next digit which is 9, storing the substring "35897" to corresponding class "3" and creating a new substring "58979" which eventually will be sent to class "5" as it happened with the first substring "53589." The process will be repeated until the length of the predefined subsequence is reached.

The final step of the first phase for creating the LERP-RSA data structure is to lexicographically sort all the substrings of the generated classes. As it can be understood, the specific step is the most time-consuming part of the analysis. In order to speed up the process, it is possible to send the different classes to be sorted in different computing devices and thus the sorting can be executed in parallel for each class reducing the overall time significantly. To estimate the time needed in the latter case, if we assume equidistribution of substrings to classes for simplifying the calculations, and assuming that we have used Classification Level 1 then for a ten-digit alphabet it easily inferred that the time needed in case of parallel sorting is approximately a little over than the one tenth of the time needed for sorting a single class or more than one hundredth if we use Classification Level 2. The sorting algorithm used is the Merge-Sort with time complexity $O(n \log n)$ which is the time complexity of the whole first phase of the method since the second step of the process mainly depends on the real-time generation or arrival of the data points in the system and thus they don't count in time complexity.

The proposed methodology continues with the second phase which involves the analysis of the previously generated classes with the ARPaD algorithm. As it

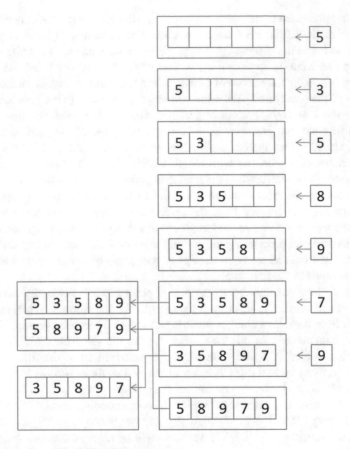

Fig. 2 The creation of the LERP-RSA data structure

was mentioned in the theoretical background, the algorithm can split the workload in several threads and analyze each class in parallel. Similarly, as in the sorting process, the classification level is very important in this phase, since the number of classes used defines how many threads will be created in the parallel execution of the algorithm and, therefore, the overall time cost. The specific phase is the second most time-consuming step of the proposed methodology since the ARPaD algorithm on average is faster than the sorting process, with average time complexity $O(n)$. Another step that may delay the algorithm is the necessity to store the results on disk for further analysis, but again the delay added is actually insignificant as it can be observed in the experimental analysis section. Depending on the hardware configuration, the specific overhead can be minimized, for example, by using enough memory to store the results on memory for meta-analyses.

The final phase of the proposed methodology is the meta-analysis. The LERP-RSA data structure has already stored all the detected repeated patterns found by the ARPaD algorithm and thus several predetermined scenarios can be run to mine valuable knowledge. The specific process is extremely fast since it actually requires the execution of a simple predefined query in the results and does not require any further processing. Furthermore, meta-analysis can also be executed in parallel or run in a separate system while the initial computer system continues to further collect and analyze the data stream using the LERP-RSA and executing the ARPaD algorithm.

3.2 Expansion of Methodology

The nature of the data structure used and the attributes of the corresponding algorithm allow more sophisticated and advanced techniques to be implemented. Such an advanced technique is presented in [21] where dynamic classification is implemented. The specific technique runs a pre-statistical analysis when an adequate number of data points have been inserted in the system and change dynamically the classification level for specific classes. For example, when 15% of the subsequence has already been entered into the system we can check the size of the classes and based on the results we can change the classification level. If we determine that one or two classes contain the majority of the data points, then we can separately change the classification level of the specific classes and create more but smaller classes. Let's assume that classification level 1 is used for the ten digits and that when 15% of the data points have been entered into the system we run the statistical analysis and find that class "9" has 65% of the suffix strings while all other classes are equally distributed with approximately 4% each one. This means that the sorting process and correspondingly the ARPaD execution in class "9" will be significantly slower than the other classes and as a consequence the overall execution time will increase considerably. However, based on the statistical analysis, the classification level for class "9" can be changed from one to two and create another ten classes instead of class "9" each one with approximately 5–6% of the observations. As a result, there will be 19 classes instead of ten classes with approximately the same size and thus the sorting and the ARPaD execution can run in parallel taking approximately the same time for all the classes and overall improved performance.

3.3 Dynamic Execution

Depending on the nature of the data and the analysis we want to perform, we can also use the dynamic sliding window approach [2]. Instead of having continuous, fixed-length, subsequence, we can use a variable size for the subsequences, for example, data points collected per time span. Furthermore, this approach allows overlapping of the subsequences in case this is important for our analysis.

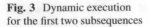

Fig. 3 Dynamic execution for the first two subsequences

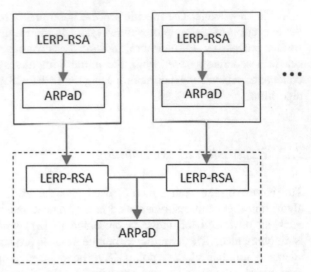

For the purpose of such analysis, we have to work with shorter subsequences and, therefore, shorter time spans. We must predefine a specific length for our subsequences based on the available hardware in order to be able to perform the analysis on memory. This is very important because we need to maximize the performance of LERP-RSA and ARPaD between the time limits that the sliding window sets. The whole process can be described in the following steps for a sliding window of size two:

1. A subsequence of predetermined length s enters the system and it is stored on memory as a LERP-RSA data structure
2. LERP-RSA is sorted
3. ARPaD is executed on the LERP-RSA and the results are examined

This is the execution process for the first subsequence that enters the system. After the first execution and having stored on memory the LERP-RSA, the full process can be described as follows (Figs. 3, 4):

1. A new subsequence of predetermined length s enters the system and it is stored on memory as a new LERP-RSA data structure
2. The new LERP-RSA is sorted
3. ARPaD is executed on the new LERP-RSA and the results are examined
4. The new LERP-RSA after step (b) is merged with the old LERP-RSA
5. ARPaD is executed on the merged LERP-RSA
6. The merged LERP-RSA is disposed from memory together with the oldest existing LERP-RSA

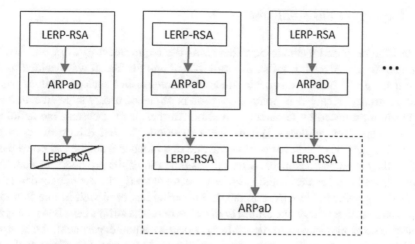

Fig. 4 Dynamic sliding window execution

It must be mentioned that (3), (1), and (2) for any new subsequence entrance are executed in parallel with steps (4) and (5). Although it seems that this is impossible, yet, such parallelism is very easy to be implemented due to the speed of the ARPaD algorithm and the merging process. More accurately, the merging process is just $O(n)$ compared to the $O(n \log n)$ sorting of the new LERP-RSA, since both previously constructed LERP-RSA and the new LERP-RSA are already sorted. Therefore, a merging process produces directly a newly sorted LERP-RSA data structure. After this, ARPaD has to be executed on the new LERP-RSA which represents a longer subsequence and, therefore, a longer time span. Although the new LERP-RSA is larger (double size in this case), yet, ARPaD can analyze it before a new LERP-RSA is created during steps (a) and (b). ARPaD is significantly faster than a sorting algorithm as it has been proved by the experimental analyses.

By executing the above-mentioned process, we have the ability to perform the following: (1) analyzing per cycle a subsequence of fixed length s which represents a time span t and (2) analyzing per cycle a longer subsequence of length $c * s$ representing a time span $c * t$ for a c number of cycles. Therefore, we do not just analyze the data points entered into the system in the last t time span but we can also analyze the new data points with the previous data points back to a specific time point of length $c * t$.

This type of execution is much more flexible than the sequential analysis described in Sect. 3.1 because it allows us to compare the newly entered data points in the system with previously entered data points and, therefore, conduct a historical analysis and comparison.

4 Experimental Analysis

To test the proposed methodology, there was the requirement to work with a very big data stream. Since it is rather difficult to find one for free, it was decided to use a composite one that have specific inherent characteristics such as randomness in the appearance of each data point, a convenient alphabet, a very large size, and can be easily reproduced by researchers to either further test the current methodology or for comparison purposes. Having these in mind, the first trillion digits of the decimal expansion of π have been chosen and downloaded from the Calico website, before they become unavailable after the termination of the provided service. Two experiments have been executed in order to test and verify the previously described methodology. For the first experiment, the π string has been split in one thousand continuous strings of size one billion digits each one. The strings have been analyzed in sequence as they appear in the π string. For the second experiment, the π string has been split in one million continuous strings of size one million digits each one. The strings have been analyzed using the dynamic execution with a sliding window of three cycles, although any kind of the number of cycles can be used. The only limitation for the number of cycles is only the available memory and CPU speed something that can be analyzed and predetermined in advance in order to achieve the best possible combination.

The experiments were run in a lab equipped with ten standard personal computers. Their hardware configuration is constituted by a quad-core Intel i7 CPU at 3.4 GHz with two threads per core (eight logical cores), 8 GB of RAM, and 400 GB of hard disk space running 64-bit operating system. For the first experiment, Microsoft SQL Server 2012 DBMS has been used to support the analysis since it is impossible to construct directly on memory datasets of sizes several gigabytes and analyze them. However, for the second experiment, every LERP-RSA process has been performed directly on memory due to the much smaller size of subsequences of length which was just one million digits for the original subsequence and 16 million for the combined sliding window.

For the first experiment, each computer executed one thousand experiments in full parallel mode using Classification Level 2 (100 classes in total) with alphabet of the digits "0" to "9," with average execution time 20 min for the LERP-RSA construction and 13 min for the ARPaD execution, adding up to 33 min (Table 1). Since a subsequence of one billion digits was analyzed in 33 min, then the maximum data stream rate could be one new digit every approximately 2 μs.

For the second experiment, each computer executed in total almost two million experiments in full parallel mode using Classification Level 2 (100 classes in total)

Table 1 Average execution time per phase

	Average time (s)	
Phase	1M	1B
LERP-RSA sorting	0.211	1188.66
ARPaD execution	0.094	761.07
Total	0.305	1949.73

Table 2 Average execution time per phase for merged subsequences

	Average time (s)				
Phase	2M	3M	4M	8M	16M
Merging	0.032	0.049	0.065	0.139	0.294
ARPaD	0.178	0.253	0.327	0.685	1.405
Total	0.210	0.301	0.392	0.824	1.699

using again the decimal system. One million experiments have been executed for the one million digits that entered the system each time. In addition, when the first three one million digits subsequences have entered the system, they have been combined to a merged LERP-RSA and then analyzed. After the analysis, the oldest LERP-RSA of the group of three subsequences is removed and when a new one million digits subsequence entered the system is merged with the two preexisted. This process continues until the end of the data stream leading to an additional, almost, one million subsequences analyses.

For each execution on the one million digits length subsequences, we have on average 0.211 s for LERP-RSA creation and sorting while we need on average 0.094 for the ARPaD algorithm execution (Table 1). This leads to a total average time of 0.305 s per million digits or data points which is equivalent to having one data point entering the system every approximately 300 ns. Although the performance of one data point every 2 μs of the first experiment is amazing, by using smaller subsequences and being able to perform all analyses on memory we managed to achieve an average execution time which is extraordinary.

The analysis of each cycle of groups of three subsequences of size one million digits each one (three in total) needs on average 0.301 s which is almost the same to the analysis of one million digits subsequences (Table 2). Therefore, for the specific hardware and under the condition that one million digits (data points) enter the system every 0.305 s, the best cycle size is three. However, this can be altered depending on how fast we have data points entering the system and the available hardware. For example, if we need to analyze one million data points per 2 s, then we are able to go up to a cycle of size 16 (Table 2) for the specific hardware. For real-world problems such weather data, financial data, etc., where data points are entering significantly slower, the size of the cycle can be analogous larger and, therefore, it will be able to analyze a much wider historical window. Regarding hardware, the more memory and processors we have, the faster we can perform the analysis because we can work on larger Classification Level, for example, Classification Level 3.

Comparing the two experiments, it has to be mentioned that the use of smaller subsequences does not produce the exact same results as the first experiment. This is obvious since in the first case, patterns that exist at the early part and the later part of the subsequence have been detected, which is not possible in the second case because simply we perform analysis at most on 16 million data points subsequences compared to one billion of the first case. Yet, this is not a drawback of the method. Analyzing data streams of few million data points in real-world cases is already a big dataset. For example, in weather data, financial data, sensor monitoring, etc., usually we can have data streams of few thousand data points per minute, compared to the million analyzed per second. This allows us to perform analysis on more complex, multivariate systems, for example, analyzing in parallel data from multiple traffic control sensors or retail store supply chain control. In such a case, we can perform a multivariate analysis on multiple data streams and be able to detect correlations between different sequences or time series, etc. [2] and provide more advanced analytics.

The presented methodology permits to perform an extremely fast and in-depth pattern detection of a data stream and afterward focus more on the meta-analyses of the results dataset. This is important since the results dataset incorporates important and meaningful knowledge which is the final target of any data streaming analytics.

5   Conclusion

Realizing the gap in the literature for an efficient and effective algorithm with an associated data structure to cope with big data streams, this chapter included an interesting methodology which satisfies the need and somehow fills the gap. The described methodology is capable of analyzing extremely big data streams using limited computing resources. It can handle both static and dynamic data streams in real time, and hence meets expectations of decision makers who are eager for timely knowledge discovery that allows them to succeed in a very competitive environment where all parties try to maximize the benefit from technology and play their own cards to lead. For this purpose, we have used an improved and customized version of our well-tested and justified methodology described in [24]. Indeed, combining the flexibility of LERP-RSA data structure and the efficiency of the ARPaD algorithm makes it possible to swiftly detect all repeated patterns, even faster than the data generation itself. This of course well satisfies the needs for dealing with dynamic data streams online and real time. Another attractive advantage of the proposed methodology is its capability to perform meta-analysis of the result by issuing advanced and complex queries directly on the results. This also allows the comparative study of time frames and the detection of useful information, for example, trends for forecasting purposes. Applicability, effectiveness, and efficiency have been well illustrated by running experiments which utilized the first one trillion digits of π. Direct usage of memory to host the data structure and for the analysis of the incoming traffic has further improved the already fast analysis of the methodology [24].

The proposed methodology is not directly comparable to other methods because it does not detect specific patterns defined by the user but, on the contrary, it detects all repeated patterns. Furthermore, the dataset used for experimental purposes is significantly larger than anything found in literature and used by other approaches. Finally, to the best of our knowledge, real-time analysis of big data streams as described in this chapter is something that has not been encountered in literature so far.

References

1. Xylogiannopoulos, K.F., Karampelas, P., Alhajj, R.: Repeated patterns detection in big data using classification and parallelism on LERP reduced suffix arrays. Appl. Intell. **45**(3), 567–597 (2016). https://doi.org/10.1007/s10489-016-0766-2
2. Xylogiannopoulos, K. F.: Data structures, algorithms and applications for big data analytics: single, multiple and all repeated patterns detection in discrete sequences. Unpublished PhD thesis, University of Calgary (2017)
3. Xylogiannopoulos, K.F., Karampelas, P., Alhajj, R.: Analyzing very large time series using suffix arrays. Appl. Intell. **41**(3), 941–955 (2014). https://doi.org/10.1007/s10489-014-0553-x
4. Apostolico, A., Preparata, F.P.: Optimal off-line detection of repetitions in a string. Theor. Comput. Sci. **22**, 297–315 (1983)
5. Weiner, P.: Linear pattern matching algorithms. In: SWAT '73 Proceedings of the 14th Annual Symposium on Switching and Automata Theory (Swat 1973), pp. 1–11 (1973)
6. Guo, D., Hu, X., Xie, F., Wu, X.: Pattern matching with wildcards and gap-length constraints based on a centrality-degree graph. Appl. Intell. **39**, 57–74 (2013)
7. Wu, Y., Wang, L., Ren, J., Ding, W., Wu, X.: Mining sequential patterns with periodic wildcards. Appl. Intell. **41**, 99–116 (2014)
8. Manber, U., Myers, G.: Suffix arrays: a new method for on-line string searches. In: Proceedings of the First Annual ACM-SIAM Symposium on Discrete Algorithms, pp. 319–327 (1990)
9. Franek, F., Smyth, W.F., Tang, Y.: Computing all repeats using suffix arrays. JALC. **8**(4), 579–591 (2003)
10. Puglishi, S.J., Smyth, W.F., Yusufu, M.: Fast optimal algorithms for computing all the repeats in a string. In: Proceedings of PSC, pp. 161–169 (2008)
11. Cormode, G., Hadjicleftheriou, M.: Methods for finding frequent items in data streams. VLDB J. **19**(1), 3–20 (2009). https://doi.org/10.1007/s00778-009-0172-z
12. Boyer, R.S., Moore, J.: A fast majority vote algorithm. Technical Report ICSCA-CMP-32, Institute for Computer Science, University of Texas (1981)
13. Demaine, E., López-Ortiz, A., Munro, J.I.: Frequency estimation of internet packet streams with limited space. In: European Symposium on Algorithms (ESA) (2002)
14. Karp, R., Papadimitriou, C., Shenker, S.: A simple algorithm for finding frequent elements in sets and bags. ACM Trans. Database Syst. **28**, 51–55 (2003)
15. Manku, G., Motwani, R.: Approximate frequency counts over data streams. In: International Conference on Very Large Data Bases, pp. 346–357 (2002)
16. Metwally, A., Agrawal, D., Abbadi, A.E.: Efficient computation of frequent and top-k elements in data streams. In: International Conference on Database Theory (2005)
17. Greenwald, M., Khanna, S.: Space-efficient online computation of quantile summaries. In: ACM SIGMOD International Conference on Management of Data (2001)
18. Shrivastava, N., Buragohain, C., Agrawal, D., Suri, S.: Medians and beyond: new aggregation techniques for sensor networks. In: Proceedings of the 2nd International Conference on Embedded Networked Sensor Systems, pp. 239–249. ACM (2004)
19. Alon, N., Matias, Y., Szegedy, M.: The space complexity of approximating the frequency moments. J. Comput. Syst. Sci. **58**, 137–147 (1999)
20. Cormode, G., Muthukrishnan, S.: An improved data stream summary: the count-min sketch and its applications. J. Algorithm. **55**(1), 58–75 (2005)
21. Xylogiannopoulos, K.F., Karampelas, P., Alhajj, R.: Sequential all frequent Itemsets detection – a method to detect all frequent sequential itemsets using LERP–reduced suffix array data structure and ARPaD algorithhm. In: Proceedings of International Conference on Advances in Social Networks Analysis and Mining, pp. 1141–1148 (2015). https://doi.org/10.1145/2808797.2809301
22. Xylogiannopoulos, K.F., Karampelas, P., Alhajj, R.: Real time early warning DDoS attack detection. In: Proceedings of the 11th International Conference on Cyber Warfare and Security, (2016), pp. 344–351 (2016)

23. Xylogiannopoulos, K.F., Karampelas, P., Alhajj, R.: Pattern detection and analysis in financial time series using suffix arrays. In: Doumpos, M., Zopounidis, C., Pardalos, P.M. (eds.) Financial Decision Making Using Computational Intelligence, pp. 129–157 (2012). https://doi.org/10.1007/978-1-4614-3773-4_5
24. Xylogiannopoulos, K.F., Karampelas, P., Alhajj, R.: Frequent and non-frequent pattern detection in big data streams: an experimental simulation in 1 trillion data points. In: Advances in Social Networks Analysis and Mining (ASONAM), pp. 931–938 (2016). https://doi.org/10.1109/ASONAM.2016.7752351

Community-Based Recommendation for Cold-Start Problem: A Case Study of Reciprocal Online Dating Recommendation

Mo Yu, Xiaolong (Luke) Zhang, Dongwon Lee, and Derek Kreager

Abstract Online dating services often use recommender systems to help people find their dates. When recommending dates to existing users who have already interacted with other users, such recommender systems tend to work well. However, recommending dates to new users who have made few interactions with others yet, the so-called "cold start" problem, still poses a problem. To address this challenge, in this paper, we propose a novel community-based recommendation framework (CBR) that can recommend dates for new users better. By detecting communities to which existing users belong and matching new users to these communities, our method is able to recommend existing users who are more likely to reply a date request from new users. Empirical validation using real data from a popular US online dating site reveals that our reciprocal online dating recommendations are significantly better than other traditional methods, achieving 5–100% improvements on average in different evaluation metrics.

Keywords Online dating recommendation · Recommender system · Cold-start · Social network analysis · Reciprocal recommendation

1 Introduction

Online dating has been widely accepted as a popular venue for finding romantic partners in recent years. According to [25], 15% of US adults have used online dating sites or mobile apps. Compared with offline dating, online dating provides

M. Yu (✉) · X. (Luke) Zhang · D. Lee
College of Information Sciences and Technology, The Pennsylvania State University, State College, PA, USA
e-mail: muy145@ist.psu.edu; lzhang@ist.psu.edu; dongwon@psu.edu

D. Kreager
Department of Sociology and Criminology, The Pennsylvania State University, State College, PA, USA
e-mail: dkreager@psu.edu

© Springer International Publishing AG, part of Springer Nature 2018
M. Kaya et al. (eds.), *Social Network Based Big Data Analysis and Applications*,
Lecture Notes in Social Networks, https://doi.org/10.1007/978-3-319-78196-9_10

more possibilities by accessing to much larger pools of candidates. At the same time, compared with offline dating, online dating users usually do not have much knowledge about users on the other end of the Internet, so sometimes it is difficult for them to identify desired dating partners from many candidates. To help their users, many online dating sites provide suggestions on potential dating partners with recommender systems.

Recommender systems have been widely studied and used over decades. A common type is based on user–item relationship, as seen in product recommendation in Amazon and movie recommendation in Netflix. In this type of recommender systems, however, the item end is passive and cannot make any choices among users. Another type is based on user–user relationship, and is widely used in online social networks such as Facebook and LinkedIn, in which a connection is reciprocal and can only be established if both ends agree. In online dating, reciprocity means that a user sends a message to another one, and then gets the other one's reply. Recommender systems for online dating need to take reciprocity into consideration, and suggest those users who are likely to reply one's dating request.

For most recommender systems, a main challenge is the cold-start problem, which refers to the issue that the system is unable to make any inferences for new items or users due to having insufficient information about them. When making recommendations for users, recommender systems usually require a large amount of information to learn preferences and interests of users, and most of such information is hidden behind user activities. Usually, the more information a recommender system can extract from a user, the better recommendations it can provide. However, for a new user without previous activities, information about the user is limited and thus it is difficult for recommender systems to accurately analyze the preferences and interests of the user. Online dating recommendation faces the same situation. It is critical to provide good recommendations to new users, because a recommender system that suggests too many undesired or non-replying connections would discourage new users to stay. A good recommender system will accelerate the process of getting new users involved in the online dating world.

With reciprocity and the cold-start problem in mind, we propose the following research question:

- Given a new online dating user who just joins online dating and has no activity, how to recommend existing users who have had online dating activities to establish promising reciprocal connections (e.g., recommended users are more likely to reply the new user's requests)?

To address this problem, we propose a novel community-based recommendation framework (CBR). We take a hybrid approach of combining ideas from both content-based and collaborative filtering recommender systems. We first identify different communities based on activities of existing users. For each community, we create its profile based on profiles of all its members. For a new user, we find the user's similarities with all communities based on profile, and then recommend existing users based on reciprocal activities of members of these communities. Using real-world data from a popular US online dating site, we test the effectiveness of our approach by comparing with existing models. Experiments show that our proposed method achieves significant improvements over existing models.

We claim the following contributions for our research. First, we propose a framework to solve the cold-start problem in online dating recommendation by combining collaborative filtering and content-based recommender systems. Second, our framework can provide reciprocal recommendations by leveraging the reciprocal contacts of existing users to generate recommendations for new users. Third, we introduce a novel idea of incorporating community detection into online dating recommendation.

In our previous ASONAM paper [33], we demonstrated effectiveness of our approach. In this paper, we further improve our previous work in the following aspects. First, we redesign the ranking scores for reciprocal contacts, and achieve better results. Second, we expand our methodology to create a set of community-based recommendation models, and compare their performance. Third, we include more datasets for experiments. With these improvements, our research is more comprehensive.

The rest of our paper is organized as follows. Section 2 introduces related work about online dating analyses as well as recommendation. Section 3 covers technical details of our framework. Section 4 describes our evaluation methods and results. We discuss our system design and performance in Sect. 5, and conclude our work with possibilities of future work in Sect. 6.

2 Related Work

Our work is inspired by some recent research on online dating. Among these works, we are more interested in literature on analyzing user behaviors and preferences, the design of recommender systems, and the cold-start problem, which are the focuses in this section.

2.1 User Behaviors and Preferences

Some researchers studied user preferences and behaviors in online dating environment. A noticeable phenomenon in online dating is homophily. Hitsch et al. [12] pointed out that users from the same race tend to match with each other. Lin and Lundquist [17] also confirmed that race plays an important role in mate selection of online dating, and in addition to homophily there exists a racial hierarchy in the reciprocating process. Skopek et al. [24] studied educational homophily in online dating.

Research also showed differences in mate selection patterns for male and female online dating users. Hitsch et al. [12] found that female users pay more attention to the income of their partners, while male users emphasize more on physical attractiveness of females. Hancock et al. [11] studied lying in online dating, and found that male users tend to lie about height while female users tend to lie about weight. This finding offers another explanation for gender differences. Kreager et al. [14] identified significant different interaction patterns for male and female users.

Some other research analyzed online dating from other angles. Fiore et al. [9] assessed attractiveness in online dating profiles, and identified photos and free-text component as two most prominent components. Xia et al. [27] conducted research about user behaviors and preferences for a famous online dating site in China. Their research pointed out that males prefer younger females while females emphasize more on socioeconomic status, and also geographic distance and photo count of users play important roles in online dating.

2.2 Online Dating Recommendation

A main stream of research about online dating recommendation relies on implicit user preferences. Akehurst et al. [2] examined explicit and implicit user preferences in online dating, and argued that explicit preferences are poor indicators of the real preferences of users. This finding was confirmed by Gemmis et al. [10]. Another work from Pizzato et al. [21] learned user implicit preferences based on activities, and took advantage of such preferences to build a system called RECON [22], one of the most famous models for reciprocal online dating recommendations. A further work from them extended RECON to consider both positive and negative preferences [23]. Chen and Nayak [5] designed a memory-based model to handle implicit preferences, and their idea was similar to that of term frequency-inverse document frequency (TF-IDF).

Another stream of work in online dating recommendation is collaborative filtering. In our early works [32, 34], we introduced a model based on user taste and attractiveness to provide reciprocal recommendations. Xia et al. [29] compared the power of several different models in providing reciprocal recommendations. Akehurst et al. [1] designed a new method to capture profile similarities, and weighted collaborative filtering based on these similarities.

Some works tried to tackle online dating recommendation from other angles. A series of works was introduced by Chen et al. In [4], they categorized users into different groups based on their activity levels, and built a customized model for each group. Another attempt by them [6] took the idea of content-based systems to cluster users into different groups based on profile similarities. For each group, they calculated user similarities, and then adopted a collaborative filtering approach to provide recommendations. In [26] Tu et al. viewed online dating as a two-sided market, and transformed online dating recommendation as an optimization problem. They built an LDA model to learn user preferences and types to match these users. Xia et al. [29] extracted several groups of features from online dating network, and compared their performances for predicting reciprocal contacts over different models. Diaz et al. [8] shaped online dating recommendation as an information retrieval problem, and tried to match users based on profiles.

2.3   Cold-Start Problem

Various methods have been proposed to address the cold-start problem. The vector-based model by Lam et al. [15] considered user information to address the cold-start problem in user-item recommendation. This model aims at finding underlying distribution over user types. However, this method only considered a limited number of combined profile attribute values. For data with more complicated user profiles, the method may suffer from data sparsity and low computational efficiency. The approach by Pereira et al. [20] clustered existing users into different groups, and used classifiers to fit new users into different clusters and then provide recommendations based on cluster profile information. One of the problems of this approach is to determine the number of clusters appropriately. Researchers also explored other different solutions by combining such techniques as classification, similarity, as well as clustering [3, 16, 31].

Some research has tackled the cold-start problem in the context of reciprocal online dating recommendation. The hybrid approach by Akehurst et al. [1] used a similarity function to identify similar existing users for a new user, and then provide the new user with recommendations based on their activities. However, the similarity function is based on the profile data of only a limited set of similar users, so overlooking much information from other users, this method may be ineffective on complicated scenarios. Also, reciprocity is not well considered in this approach. Kim et al. [13] proposed an approach of suggesting recommendations based on user subgroups identified based on profile attributes, but this method may also suffer from drawbacks of content based methods, such as the difficulty of defining customized similarity functions. Another classical work is RECON [22], which pairs a new user with existing users who will be interested in the new user. However, its recommendation is purely based on the preferences of existing users and takes no reciprocity into consideration.

3   Methods

In this section we describe our CBR framework in detail. Our approach combines content-based and collaborative filtering recommender systems. For new users, we find their similarities to existing users, and then use the reciprocal activities of those existing users to provide recommendations. Instead of calculating similarities between new users and every existing user, we group existing users into different communities, and then compare new users with those communities.

As shown in Fig. 1, our framework includes five key steps. The first two steps are to extract communities from existing users, and find associated reciprocal contacts with each community. The third step is to generate community profiles. New users are matched to these communities in the fourth step. The final step is to make recommendations.

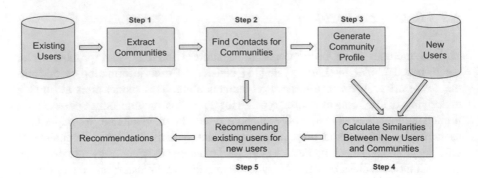

Fig. 1 Workflows of proposed recommender system

3.1 Extracting Communities from Existing Users

The first step in our framework is to extract communities from existing users. The goal of this step is to find different groups of similar users. Community detection is well studied in the fields of machine learning and social network analyses, and various methods have been proposed [7, 18, 19]. In these methods, the input is a social network, and the output are several subnetworks of the network.

We generate an input social network for community detection algorithm as follows. First, we separate existing male and female users, and generate two social networks for them separately. Here we use male user network as an example, and female user network is constructed in the same fashion. Let $G = \{V, E\}$ be the male user similarity network. V is a set of vertices representing existing male users. We define $E = \{e_{x,y}, x \in V, y \in V, x \neq y\}$ as the set of weighted edges. The weight of an edge represents similarity between the pair of users on two ends. For any male user $x \in V$, let C_x be the set of female users who have contacts with him. Here we introduce three ways of defining C_x, following similar fashion in [32]. The first definition is the set of female users who were contacted by x. The second definition is the set of female users who were contacted by x or who contacted x and got his replies. The third definition is the set of users who were contacted by and also replied to x. An example of generating C_x is shown in Fig. 2. Given a male user Bob, under the first definition, {Alice, Michelle} will be his C_x, and under the second and third definitions, his C_x will be {Alice, Michelle, Jenny} and {Alice} separately. These three definitions evaluate user activities from different aspects, and we will test their effectiveness in experiments.

We can also define C_x as the set of users who have contacted x, but such definition contradicts our idea of extracting communities based on implicit user preferences. For example, for two users A and B, if C_A and C_B are the same, then it means that A and B are favored by the same group of opposite gender users. A and B will probably have very similar user profiles. When we detect communities for existing users and then match new users to these communities, we are actually matching new users to existing users based on profile similarity. Similar work has already been reported in [1].

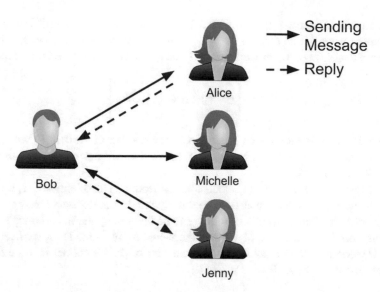

Fig. 2 An example of generating C_x. Under three definitions, C_x for Bob is {Alice, Michelle}, {Alice, Michelle, Jenny}, and {Alice} separately

After defining C_x, we define the weight of $e_{x,y}$ as:

$$w_{e_{x,y}} = \frac{|C_x \cap C_y|}{|C_x \cup C_y|}$$

$w_{e_{x,y}}$ is the Jaccard coefficient between C_x and C_y, which measures the overlap between $x's$ and $y's$ contacts. The more common contacts they share, the more similar two users are, and the higher weight there will be. We also follow the same fashion of [30] to add a prior similarity score to $w_{e_{x,y}}$ if two users do not share any common neighbors. Under such definition, we find weights for all edges in E.

After getting G, we run community detection algorithm over it. Community detection on large networks is a challenging task, because many algorithms run very slowly even for networks with a few thousand nodes. Here we use the algorithm introduced in [7], considering its speed and overall performance. The output of community detection algorithm is a set of communities T. For each community $t \in T$, $t = \{V_t, E_t\}$, where $V_t \subset V$ is the set of users who belong to t, and $E_t \subset E$ is the set of edges among all users in V_t.

3.2 Finding Reciprocal Contacts for Communities

With T detected in the previous step, we can find reciprocal contacts for these communities.

For each community $t \in T$, we have V_t as the set of users who belong to t. For a male user x, we define CR_x as the set of female users who were contacted by and also replied to x. Then we create a distribution R_t for t, which is defined as follows:

$$R_t = \left\{ (u, m) : u \in \bigcup_{x \in V_t} CR_x \right\}$$

where m is response rate, which is defined by the number of users in V_t who got u's replies to the number of users in V_t who contacted u. We normalize m with regard to the total of all ms in R_t.

For a female user, the more contacts she got from and also replied to t, the more likely she was favored by and also favored users in t. For male users who are similar to those in t, this female user is also likely to be favored by them, and once she gets contacts from them, it is also likely that she would reply them. Thus, we keep the n here for recommendation ranking purpose in later steps. We collect R_t for all $t \in T$ and denote this set as R, $R = \{R_t, t \in T\}$.

3.3   Generating Community Profile

Each user has a profile. We also want to create a profile for each community. Here we follow the same fashion to find user preferences in [22]. In user profiles, there are several attributes, such as race, height, and body type. We define the list of attributes as A. We also define the profile of a user x as

$$U_x = \{val_a : \text{for all attributes } a \in A\}$$

where val_a is a value of attribute a.

For community t, attribute a in its profile is represented as $p_{t,a}$, which is defined as follows:

$$p_{t,a} = \{(val, n) : \text{for all unique discrete values } val \text{ of } a\}$$

where n is the number of times val occurred in V_t. Note that because values in a are discrete, $p_{t,a}$ prefers categorical values.

The profile P_t for community t can then be represented as:

$$P_t = \{p_{t,a} : \text{for all } a \in A\}$$

The concept of community profile is similar to user preferences defined in [22]. An example of generating community profile is shown in Fig. 3. In general, the profile of a community is represented as a set of distributions, and each distribution is corresponding to an attribute, showing the number of times each discrete value has occurred. We collect profiles for all communities, and denote the set as P, $P = \{P_t, t \in T\}$.

Fig. 3 An example of generating community profile. Here we use a community with six users, and two attributes

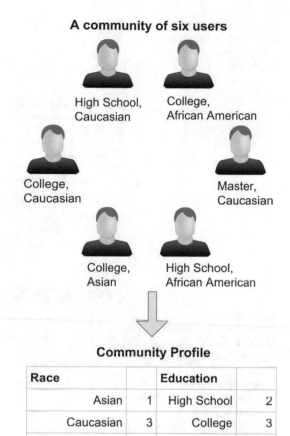

A community of six users

High School, Caucasian

College, African American

College, Caucasian

Master, Caucasian

College, Asian

High School, African American

Community Profile

Race			Education	
Asian	1	High School	2
Caucasian	3	College	3
African American	2	Master	1

3.4 Calculating Similarities Between New Users and Communities

For each new user, we calculate the similarity scores between the user and all communities. We still take the similar fashion of fitting a user to another's preferences introduced in [22], but here we fit a user x to the profile of each community t, and then get a similarity score between x and t. We denote the score as $Sim(x, t)$. We normalize $Sim(x, t)$ with regard to the total of all $Sim(x, t)$ for $t \in T$. Details are in Algorithm 1. An example is shown in Fig. 4.

The similarity score between a new user and a community measures to what extent the new user may belong to the community. Although our communities are detected based on user activities, we cannot infer a new user's association with a community based on activities, because there are no observations. Thus, we take the idea from content-based recommender systems, and then calculate similarities between new user profile and community profiles. These similarity scores will be used for the final step of recommendation.

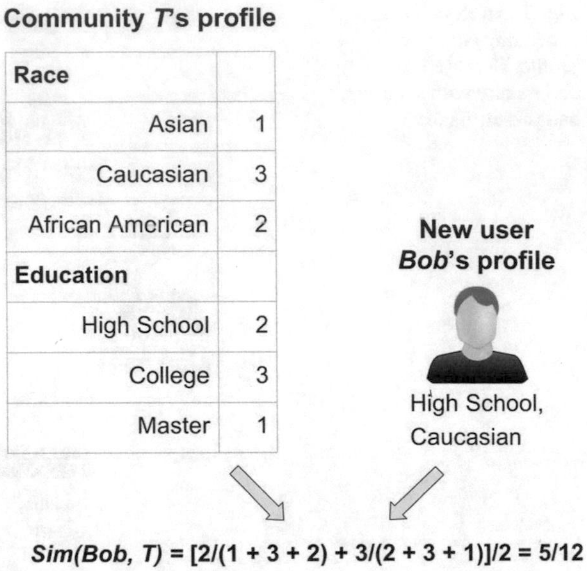

Fig. 4 An example of calculating a new user's similarity score to a community

Community *T*'s profile

Race	
Asian	1
Caucasian	3
African American	2
Education	
High School	2
College	3
Master	1

**New user
Bob's profile**

High School,
Caucasian

$Sim(Bob, T) = [2/(1 + 3 + 2) + 3/(2 + 3 + 1)]/2 = 5/12$

Algorithm 1 Calculate $Sim(x, t)$

Input: User x, Community profile P_t
Output: Similarity score $Sim(x, t)$
  $s = 0$
  for each $a \in A$ **do**
    Get the value v_a of attribute a in U_x
    Get $p_{t,a}$ in P_t
    Find $(v, n) : (v, n) \in p_{t,a}, v = v_a$
    if $n! = 0$ **then**
      $s \leftarrow s + \frac{n}{\Sigma n \in p_{t,a}}$
    end if
  end for
  return $\frac{s}{|A|}$

3.5 Recommendation

Our goal in this step is to recommend existing users to new users, with the hope of forming reciprocal contacts between them. To recommend an existing user x to a new user y, we find a recommending score $Recom(x, y)$ between them. The algorithm of finding the recommending score is described in Algorithm 2.

For a new user y, we first find the similarities between y and all communities. This step is to find those similar users that y can learn from. For a community t, we check if they have reciprocal contacts with an existing user x, and get the value m, i.e. response rate, from the pair (x, m). We calculate product of $Sim(y, t)$ and m, and take the product as the score of recommending x to y based on information of the community t. We do such calculations over all communities, so the user

Algorithm 2 Calculate $Recom(x, y)$

Input:
  Existing user x, New user y, Community profile P,
  Reciprocal contact distribution R
Output: Recommending score $Recom(x, y)$
  $RScore = 0$
  for each $t \in T$ **do**
    Calculate $s = sim(y, P_t)$
    Get (x, m) if $(x, m) \in R_t$
    $RScore \leftarrow RScore + s * m$
  end for
  $Recom(x, y) = RScore$
  return $Recom(x, y)$

y can learn from not just one but all communities, with different weights. We sum up recommending scores from all communities, and take the sum as the final recommending score between x and y. We evaluate those scores between the new user y and all existing users, and generate a recommended list in a descending order of recommending scores for x. An example is shown in Fig. 5.

4 Evaluation

To evaluate effectiveness of our approach, we ran experiments on data from a major US online dating site. We used four metrics to measure performance: precision, recall, normalized discounted cumulative gain (NDCG), and mean average precision (MAP). All these metrics are commonly used to evaluate recommender systems. The first two were used in previous work of online dating recommendation [22, 28, 29], but these metrics do not consider positions of recommendations. We wanted to not only find desired dating partners for new users, but also put these partners on top of the recommended list. NDCG and MAP both take positions into consideration, and are commonly used in evaluations considering ranking.

We introduce details of our dataset as well as experiments in the following sections.

4.1 Dataset

To test the effectiveness of our approach, we ran experiments on real-world data from a major US online dating site. Our datasets contain two cities, one in southwest and another in midwest. The race compositions of these two cities are different, and one city has more minorities. Other demographics are similar. We included all users from these two cities, as well as their associated dating activities, for a time

Fig. 5 An example of recommendation for a new user *Bob*

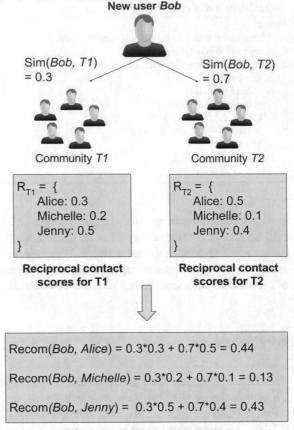

New user *Bob*

Sim(*Bob*, *T1*) = 0.3

Sim(*Bob*, *T2*) = 0.7

Community *T1*

Community *T2*

R_{T1} = {
  Alice: 0.3
  Michelle: 0.2
  Jenny: 0.5
}

R_{T2} = {
  Alice: 0.5
  Michelle: 0.1
  Jenny: 0.4
}

Reciprocal contact scores for T1

Reciprocal contact scores for T2

Recom(*Bob*, *Alice*) = 0.3*0.3 + 0.7*0.5 = 0.44

Recom(*Bob*, *Michelle*) = 0.3*0.2 + 0.7*0.1 = 0.13

Recom(*Bob*, *Jenny*) = 0.3*0.5 + 0.7*0.4 = 0.43

Recommendation results for *Bob*

span of 196 days. There are two parts in the dataset. The first part is user profile, and the second part is an activity log. A user profile describes basic demographic information, as well as some personal habits of a user. Each record in the activity log represents a contact between a pair of users. It contains sender ID, receiver ID, time stamp of their first message, as well as information of whether a receiver replied to a sender or not. All users in datasets are anonymized.

We performed profile attribute selection on our dataset. When users create their profiles, some attributes are required, and some are optional. We found that a few optional attributes were left blank by most of users. Because our approach needs to generate community profile and match new users to communities, missing values cannot provide much information and may even be misleading. Thus, we dropped three attributes with missing values for more than 50% of all users. For the remaining attributes, we selected users with no blank for attribute. Because most users put values for the remaining attributes, during this selection we lost few users (13.7%). Also, in some attributes, some values are similar, so we combined them and made only one value. We also discretized continuous attributes to create different categories. A description of user profile is in Table 1. The combined values are indicated within parentheses. The only attribute that may need explanation is quickmatch rating. On the online dating website, a user's profile will be randomly assigned to some others for rating, in a scale of 1–5. The mean value of all collected ratings is assigned as quickmatch rating for the user.

We split our data into two parts to collect training data as well as new users for experiments. The first part was to identify existing users and detect communities among them. We used data from a period of 100 days, and we called it training phase. If a user was involved in any activities, including both sending and receiving contacts, during training phase, then we identified the user as an existing user. Community detection was based on the activities of these existing users during training phase. The second part was to make recommendations for new users. We used the following 30 days after training phase to identify new users, and we called this time period as selecting phase. For any users who had sending or receiving activities in selecting phase, if they had no activities during training phase, we identified them as new users.

There were some potential issues with our approach of identifying existing and new users. We realized that the best way of identifying new and existing users is to look at their registration time. However, such information is not available in our dataset. Our method may identify some registered users who were inactive during training phase as new users. Thus we created a relatively long training period to mitigate such effect. Another potential issue is that, in the selecting phase, early new users became existing users for late new users, and there were interactions among them. To mitigate such effect, we chose a relatively short selecting phase. Our analyses confirmed that "new-to-new" interactions only represented a small proportion of new user activities. We removed these "new-to-new" interactions during our experiments. Last but not least, our dataset covered user activities for 196 days, but here training phase only required 100 days data. The reason of not using more data for training was due to our evaluation methodology. Instead of running one experiment to report results, we followed similar fashion of cross validation, and ran five experiments to conduct a more comprehensive evaluation. Note that the evaluation is not exactly based on cross validation, because our method is time dependent. From the whole dataset we chose five 100-day periods for training separately. Ideally there should be no overlap among all experiments, but because of the limitation on data size, we used relatively short training phase to minimize overlaps among experiments.

With identified new users in hand, we collected testing data for experiments. Based on our analyses, most activities of new users happened within a short period after they joined online dating. This phenomenon was also pointed out by other researchers in previous work [29]. For a new user x, we looked at the user's activities in the first 10 days, and generated CR_x for the user. As mentioned before, we removed "new-to-new" interactions, thus CR_x is a set of existing users only.

With all these preparations, we finalized training and testing datasets. A brief description of our data is shown in Tables 2 and 3 for city A and B separately.

Table 1 Description of user profile

Attribute	Value
Race	White, Black, Multiple, Other
Education level	High School, Postgrad University (University, 2YR)
Education status	Dropped, Graduated, Working On
Body type	Average, Thin (Thin, Skinny) Fit (Athletic, Fit, Jacked) Overweight (Overweight, Obese, Fleshy, Full_Figured)
Smoking	No, Sometimes, When drinking, Yes, Trying to quit
Drinking	Never, Rarely, Socially, Often, Very often, Desperately
Quickmatch rating	Bucketized, from 1 to 5
Height	Bucketized, from 0 to 9
Age	Discreted, from 3 to 23

Table 2 Dataset description of city A

Item	Experiment				
	1	2	3	4	5
Existing user	6686	6863	7052	7210	7355
Total new user	1060	1129	1094	1037	1009
New user who contacted existing users	418	442	449	444	451
Contacts sent by new users	5047	5614	5164	5066	4965
Contacts from new to existing users	79%	76%	77%	78%	79%
Reciprocal contacts sent by new users	1204	1383	1373	1367	1263
Reciprocal contacts from new to existing users	76%	73%	74%	76%	79%

Table 3 Dataset description of city B

Item	Experiment				
	1	2	3	4	5
Existing user	5468	5625	5823	6067	6276
Total new user	1043	1115	1109	1054	992
New user who contacted existing users	375	391	411	422	403
Contacts sent by new users	3936	3997	4026	3981	3886
Contacts from new to existing users	75%	72%	73%	76%	77%
Reciprocal contacts sent by new users	981	1035	1111	1064	1075
Reciprocal contacts from new to existing users	71%	66%	67%	73%	72%

4.2 Experiments

We conducted two sets of experiments. The first set was to test performance of different variations of CBR-based methods, and the second set was to compare CBR-based methods with baselines.

4.2.1 Comparing Variations of CBR

Our first set of experiments was to test performance of different variations of CBR. As described earlier, we defined three ways of constructing C_x. As a result, we developed three CBR-based models: sending similarity CBR (CBR-S), sending-and-replying similarity CBR (CBR-SR), and reciprocal similarity CBR (CBR-REC).

We calculated all metrics in top-K recommendation. Figures 6 and 7 show results based on city A and B data, respectively.

Our results showed that CBR-REC performed worst in all metrics across both cities. This may be due to the fact that reciprocal contacts are very limited in online dating (20% of all contacts in our dataset), so CBR-REC did not have sufficient data for training. For CBR-S and CBR-SR, results varied by metrics and cities. In general, the performance of CBR-S was better than CBR-SR on city A's data, with the exception of recall that both of them achieved similar results. However, for city B data, CBR-SR had better results for precision and recall. For NDCG, performance of both methods was similar. CBR-S achieved around 40% improvement against

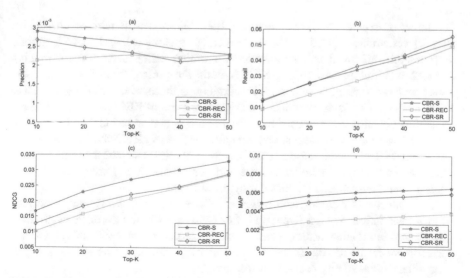

Fig. 6 Comparison of community-based recommender systems with city A data. Top-K results of (**a**) precision, (**b**) recall, (**c**) NDCG, and (**d**) MAP

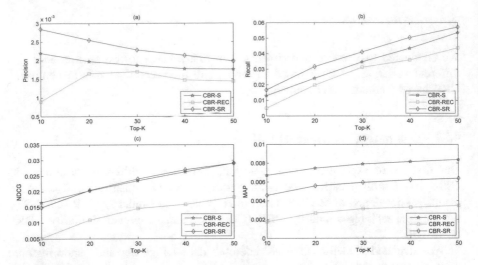

Fig. 7 Comparison of community-based recommender systems with city B data. Top-K results of (**a**) precision, (**b**) recall, (**c**) NDCG, and (**d**) MAP

CBR-SR for MAP on city B's data. With mixed results for CBR-S and CBR-SR, we decided to keep both of them for the next round of experiments to compare with other two baselines.

4.2.2 Comparing CBRs with Baselines

In this set of experiments, we compared CBR-S and CBR-SR with two baselines. In [22] researchers introduced a recommender system called RECON, aiming at providing reciprocal recommendation for online dating users. As for making recommendations to new users, RECON finds the preferences of existing users based on their sending activities, and recommends those who would be interested in the new users to them. We adopt the original version of RECON as one baseline. In another baseline we only used reciprocal contacts that were initiated by existing users to learn their preferences, and we call this model rRECON. The remaining parts of rRECON are the same as RECON. Both baselines shared the same training phase with our proposed model, and the preferences of existing users were learned based on their associated activities during training phase.

To conduct a fair comparison between different models, we followed the idea of cross-validation. We ran five rounds of experiments based on different training phases and associated testing phases, and took the mean value of all metrics for each model to report. Tables 4 and 5 show performance on different top-K recommendations of city A and B, respectively.

Experiment results clearly show that rRECON outperformed the original RECON, so we only compare our CBR-SR and CBR-S with rRECON, and show

Table 4 Experiment results for city A (10^{-3})

	TopK	RECON	rRECON	CBR-SR	CBR-S
Prec	10	1.13 [0.44]	1.63 [0.37]	2.68(64.4%) [0.41]	**2.90(77.9%) [0.43]**
	20	1.59 [0.45]	1.57 [0.34]	2.47(57.3%) [0.40]	**2.72(73.2%) [0.49]**
	30	1.29 [0.35]	1.59 [0.31]	2.34(47.2%) [0.33]	**2.62(64.8%) [0.38]**
	40	1.43 [0.36]	1.46 [0.27]	2.11(44.5%) [0.38]	**2.43(66.4%) [0.24]**
	50	1.41 [0.31]	1.46 [0.20]	2.21(51.4%) [0.27]	**2.31(58.2%) [0.24]**
Rec	10	6.25 [2.1]	9.31 [2.9]	**14.72(58.1%) [2.5]**	13.92(49.5%) [2.4]
	20	14.70 [2.9]	16.78 [2.7]	25.36(51.5%) [3.0]	**25.86(54.1%) [3.1]**
	30	18.37 [3.6]	25.71 [4.9]	**36.41(41.6%) [4.5]**	34.14(32.8%) [4.7]
	40	27.78 [5.6]	31.86 [4.5]	**43.54(35.6%) [5.4]**	42.20(32.5%) [4.5]
	50	35.33 [5.0]	38.50 [3.5]	**55.65(44.5%) [4.2]**	51.66(34.2%) [4.3]
NDCG	10	5.05 [2.2]	7.80 [1.9]	12.65(62.2%) [2.6]	**16.69(114.0%) [2.8]**
	20	10.14 [3.0]	11.67 [3.0]	18.28(56.6%) [2.7]	**22.86(95.9%) [3.1]**
	30	11.60 [3.3]	15.19 [3.5]	22.12(45.6%) [3.8]	**26.93(77.3%) [3.4]**
	40	15.02 [4.2]	17.31 [3.8]	24.72(42.8%) [5.2]	**30.11(73.9%) [4.2]**
	50	17,41 [4.5]	19.94 [3.6]	28.82(44.5%) [4.9]	**32.99(65.4%) [3.5]**
MAP	10	1.26 [0.48]	1.92 [0.52]	4.21(119.3%) [0.55]	**4.89(154.7%) [0.50]**
	20	1.77 [0.47]	2.42 [0.54]	4.96(105.0%) [0.50]	**5.69(135.1%) [0.52]**
	30	1.91 [0.51]	2.78 [0.61]	5.42(95.0%) [0.54]	**6.05(117.6%) [0.53]**
	40	2.18 [0.49]	2.95 [0.61]	5.62(90.5%) [0.54]	**6.29(113.2%) [0.55]**
	50	2.35 [0.49]	3.09 [0,59]	5.91(91.3%) [0.57]	**6.51(110.7%) [0.51]**

Bold values are best values within all methods. Significance is tested in Table 6.

the lifts of our methods in parentheses, as well as the standard deviation across results of all rounds of experiments in bracket. For city A, CBR-S achieved the best for precision, NDCG and MAP, while CBR-SR was slightly better than CBR-S for recall. Both CBR-SR and CBR-S outperformed rRECON. For city B, CBR-SR achieved the best for precision and recall. It also got the best results for NDCG from top-20 to top-40. For MAP, CBR-S outperformed the other methods greatly. Compared with rRECON, CBR-SR outperformed it for all metrics at all levels, while CBR-S showed slight drops for precision, but beat it for all other metrics. The standard deviation of all models are similar.

We performed paired t-test for CBR-S and rRECON, as well as for CBR-SR and rRECON, for all metrics. Results are shown in Table 6. CBR-SR achieved significant improvements over rRECON on all metrics for both datasets. For CBR-S, it showed no significant drop on precision for city B, but achieved significant improvements over all other metrics for both datasets.

In general both CBR-SR and CBR-S gave better results than rRECON. Experiment results validated the effectiveness of our CBR-based recommendation framework.

Table 5 Experiment results for city B (10^{-3})

	TopK	RECON	rRECON	CBR-SR	CBR-S
Prec	10	1.90 [0.42]	2.70 [0.38]	**2.85(5.6%) [0.45]**	2.20(−18.5%) [0.40]
	20	1.65 [0.40]	1.92 [0.18]	**2.55(32.8%) [0.67]**	1.97(2.6%) [0.29]
	30	1.55 [0.33]	2.05 [0.26]	**2.28(11.2%) [0.38]**	1.86(−9.3%) [0.29]
	40	1.45 [0.33]	1.92 [0.13]	**2.14(11.5%) [0.25]**	1.77(−7.8%) [0.19]
	50	1.39 [0.35]	1.81 [0.20]	**1.98(9.4%) [0.20]**	1.76(−2.8%) [0.16]
Rec	10	8.52 [2.9]	11.24 [1.5]	**16.68(48.4%) [2.8]**	12.95(15.2%) [3.1]
	20	16.38 [3.0]	19.75 [3.2]	**31.66(60.3%)[3.2]**	24.15(22.3%) [3.6]
	30	24.06 [3.5]	34.03 [5.7]	**41.00(20.5%) [4.1]**	34.64(1.8%) [4.9]
	40	30.16 [6.1]	42.16 [5.6]	**50.42(19.6%) [6.6]**	43.45(3.1%) [4.8]
	50	35.90 [7.2]	48.52 [8.3]	**57.06(17.6%) [8.2]**	53.47(10.2%) [5.3]
NDCG	10	10.21 [1.8]	14.91 [2.3]	14.95(0.3%) [2.1]	**16.53(10.9%) [2.5]**
	20	13.19 [2.8]	17.79 [2.1]	**20.46(15.0%) [3.1]**	20.35(14.4%) [2.3]
	30	15.83 [3.3]	22.41 [2.5]	**24.09(7.5%) [3.0]**	23.53(5.0%) [2.1]
	40	17.95 [4.0]	25.41 [1.7]	**27.03(6.4%) [4.7]**	26.37(3.8%) [3.8]
	50	20.03 [4.9]	27.93 [3.0]	29.09(4.2%) [3.0]	**29.14(4.3%) [4.2]**
MAP	10	2.31 [0.62]	3.55 [0.58]	4.58(29.0%) [0.57]	**6.71(89.0%) [0.63]**
	20	2.86 [0.59]	4.09 [0.58]	5.59(36.7%) [0.59]	**7.46(82.4%) [0.61]**
	30	3.17 [0.64]	4.65 [0.54]	5.95(28.0%) [0.60]	**7.90(69.9%) [0.54]**
	40	3.35 [0.61]	4.88 [0.54]	6.22(27.5%) [0.56]	**8.16(67.2%) [0.57]**
	50	3.48 [0.65]	5.01 [0.48]	6.38(27.4%) [0.59]	**8.38(67.3%) [0.54]**

Bold values are best values within all methods. Significance is tested in Table 6.

Table 6 t-Test results for CBR-SR and CBR-S over rRECON

	City A				City B			
	Precision	Recall	NDCG	MAP	Precision	Recall	NDCG	MAP
CBR-SR	***	**	***	***	*	**	*	***
CBR-S	***	**	***	***	–	*	**	***

(–) No significance; $*p<0.05$; $**p<0.01$; $***p<0.001$

5 Discussion

Our experiment results show that CBR-based approaches are better than conventional existing methods for the cold-start problem in reciprocal online dating recommendation. Now we further analyze the strengths and weaknesses of each model.

The first baseline RECON has been proven as an effective method for reciprocal online dating recommendation. It considers a user's sending activities, and analyzes the user's preferences over all available attributes, and fits each potential online dating partner to these preferences to get a compatibility score. However, the effectiveness of this method is more significant for existing users than for new users. In the original research of RECON [22], to match a pair of online dating users, the model must consider the compatibility scores of both ends. This reflects the idea of reciprocity in online dating recommendation. Under cold start scenarios, however, because new users have no activities, it is impossible to get their preferences over different attributes. As a result, the effectiveness of RECON on online dating is weak.

To overcome RECON's limitation, we consider reciprocity and introduce rRECON as another baseline. rRECON uses reciprocal contacts of a user to learn his/her preferences, and thus mitigates the problem of lacking of reciprocity in RECON. Experiments show that it achieves considerable improvement over the original model. However, a potential threat to rRECON is that most of sending contacts in online dating get no responses. In our dataset, there are only around 20% of all sending contacts finally turn into reciprocal ones. Some other research also reported similar observation, for example in [29] they reported even lower reciprocal rates (10% for male senders and 18% for female senders). Also, RECON-based models are all trained on individual basis. However, activities on individual level are very limited, so such effect may lead to even worse results. Previous works pointed out that using only reciprocal contacts for online dating recommendation may not get desired results [32, 34].

Our CBR-based methods achieved significant improvement over RECON-based methods. The success of CBR-based methods is based on the following factors.

First, we take advantages of both content-based and collaborative filtering recommender systems. We create profiles for different communities, and match new users to these communities based on profile similarity. By doing this, we take advantage of content-based methods. Also, because we follow the same fashion of RECON to create community profile as distributions over different attributes, we avoid finding customized similarity functions, and thus eliminate drawbacks of content-based methods. Unlike RECON and rRECON which overlook activities of similar users, in CBR-based methods we match new users to all communities. This step takes ideas from collaborate filtering, and ensures that new users can learn from all similar users.

Second, we introduce community detection into our design of online dating recommendation. As we stated before, finding communities enables new users to learn from all existing users. Also, the way we build input networks for community detection takes ideas from collaborative filtering. Instead of directly generating a network based on topology, we build two networks based on similarities of messaging patterns of existing male and female users separately. This method ensures that we divide existing users based on activity patterns, so new users can learn from existing users who are likely to have similar activities. We tested three ways of defining user connections for community detection. Experiments show that CBR-REC was not comparable to other two methods. This may be due to limitedness of reciprocal activities, which also constrains rRECON.

Third, our framework emphasizes on reciprocity for recommendation. Existing works usually only consider one side's preferences, because they are not able to properly infer the interests of new users. Experiment improvements from RECON to rRECON confirmed importance of reciprocal information. In our methods, we utilize reciprocal information from all communities to help new users. Through such a design, we consider the preferences of both ends, and thus properly utilize reciprocal information in recommendation.

There are certain limitations about our work. First, we only apply one community detection method, and there are other methods that we can apply and compare. Second, for each community we only retrieve reciprocal contacts, and thus miss some information from one-way contacts. Third, we adopt RECON's fashion of calculating a user's similarity score to a community, but other approaches of estimating the similarity score can be used. Fourth, our experiments are based on one major US online dating sites, more datasets will further validate effectiveness of our framework. Also, performance of CBR-based models are different for two cities. Considering demographic compositions of these two cities are different, the results may imply that our models may need further tuning for different scenarios.

In sum, RECON-based methods do not take reciprocity as well as user similarities into consideration, and thus do not provide ideal reciprocal recommendations to new users. In contrast, our CBR-based methods borrow ideas from both content-based and collaborative filtering recommender systems. We match new users to different communities of existing users, and take advantage of the reciprocal activities of community members to make recommendations for new users. Experiment results demonstrated the effectiveness of our approach.

6 Conclusion and Future Work

In this paper, we reported our research on a novel framework for recommending reciprocal contacts from existing to new online dating users. Our work borrowed ideas from both collaborative filtering and content-based recommender systems, and took advantage of both user similarities and reciprocity to make successful reciprocal online dating recommendations. Using real-world data from a popular US online dating site, our approach demonstrated significant improvement over existing methods.

Although our work achieved good results, there are spaces for further improvement. First, we are interested in knowing the impact of community detection algorithms on our results. Our current research is primarily based on a popular divisive algorithm [19]. Other community detection algorithms may lead to different community structures, and consequently could affect the recommendation results. Second, we will further improve our approach by considering information from sending contacts, which are much more than reciprocal contacts. Combining these two types of contacts could potentially lead to better results. Third, we will explore other approaches of estimating the similarities of new users to communities, and compare their effects on the performance of recommendation. Fourth, we will improve the overall performance of our framework by systematically optimizing and combining different parts in our approach. Furthermore, we will validate our method with datasets from other online dating sites, in particular those exhibiting different user profile features and community structures from those of the data used in our study. Doing so will help us better understand the generality of our method.

References

1. Akehurst, J., Koprinska, I., Yacef, K., Pizzato, L., Kay, J., Rej, T.: CCR - a content-collaborative reciprocal recommender for online dating. In: Proceedings of the Twenty-Second International Joint Conference on Artificial Intelligence, vol. 3, pp. 2199–2204 (2011). https://doi.org/10.5591/978-1-57735-516-8/IJCAI11-367
2. Akehurst, J., Koprinska, I., Yacef, K., Pizzato, L., Kay, J., Rej, T.: Explicit and implicit user preferences in online dating. In: The Behavior Informatics 2011 (BI2011) Workshop - The 15th Pacific-Asia Conference on Knowledge Discovery and Data Mining (PAKDD2011), pp. 15–27. Springer, Berlin (2011)
3. Bobadilla, J., Ortega, F., Hernando, A., Bernal, J.: A collaborative filtering approach to mitigate the new user cold start problem. Knowl.-Based Syst. **26**, 225–238 (2012)
4. Chen, L.: A recommendation approach dealing with multiple market segments. In: 2013 IEEE/WIC/ACM International Joint Conferences on Web Intelligence (WI) and Intelligent Agent Technologies (IAT), vol. 1, pp. 89–94. IEEE (2013). https://doi.org/10.1109/WI-IAT.2013.13
5. Chen, L., Nayak, R.: A reciprocal collaborative method using relevance feedback and feature importance. In: 2013 IEEE/WIC/ACM International Joint Conferences on Web Intelligence (WI) and Intelligent Agent Technologies (IAT), vol. 1, pp. 133–138. IEEE (2013). https://doi.org/10.1109/WI-IAT.2013.20
6. Chen, L., Nayak, R., Xu, Y.: A recommendation method for online dating networks based on social relations and demographic information. In: 2011 International Conference on Advances in Social Networks Analysis and Mining, pp. 407–411. IEEE (2011). https://doi.org/10.1109/ASONAM.2011.66
7. Clauset, A., Newman, M.E.J., Moore, C.: Finding community structure in very large networks. Phys. Rev. E: Stat. Nonlinear Soft Matter Phys. **70**(6 Pt 2), 1–6 (2004). https://doi.org/10.1103/PhysRevE.70.066111
8. Diaz, F., Metzler, D., Amer-Yahia, S.: Relevance and ranking in online dating systems categories and subject descriptors. In: Proceedings of the 33rd International ACM SIGIR Conference on Research And Development in Information Retrieval, pp. 66–73. ACM, New York (2010). https://doi.org/10.1145/1835449.1835463
9. Fiore, A.T., Taylor, L.S., Mendelsohn, G.A., Hearst, M.: Assessing attractiveness in online dating profiles. In: Proceeding of the Twentysixth Annual CHI Conference on Human Factors in Computing Systems CHI 08, 1992, p. 797. ACM, New York (2008). https://doi.org/10.1145/1357054.1357181. http://portal.acm.org/citation.cfm?doid=1357054.1357181
10. Gemmis, M.D., Iaquinta, L., Lops, P., Musto, C., Narducci, F., Semeraro, G.: Preference learning in recommender systems. In: Preference Learning, vol. 41 (2009)
11. Hancock, J.T., Toma, C., Ellison, N.: The truth about lying in online dating profiles. In: CHI Proceedings, pp. 449–452. ACM, New York (2007). https://doi.org/10.1145/1240624.1240697
12. Hitsch, G.J., Hortaçsu, A., Ariely, D.: What makes you click?-Mate preferences in online dating. Quant. Mark. Econ. **8**(4), 1–35 (2010). https://doi.org/10.1007/s11129-010-9088-6
13. Kim, Y.S., Krzywicki, A., Wobcke, W., Mahidadia, A., Compton, P., Cai, X., Bain, M.: Hybrid techniques to address cold start problems for people to people recommendation in social networks. In: Pacific Rim International Conference on Artificial Intelligence, pp. 206–217. Springer, Berlin (2012)
14. Kreager, D.A., Cavanagh, S.E., Yen, J., Yu, M.: "Where Have All the Good Men Gone?" Gendered interactions in online dating. J. Marriage Fam. **76**(April), 387–410 (2014). https://doi.org/10.1111/jomf.12072
15. Lam, X.N., Vu, T., Le, T.D., Duong, A.D.: Addressing cold-start problem in recommendation systems. In: Proceedings of the 2nd International Conference on Ubiquitous Information Management and Communication, pp. 208–211. ACM, New York (2008)
16. Lika, B., Kolomvatsos, K., Hadjiefthymiades, S.: Facing the cold start problem in recommender systems. Expert Syst. Appl. **41**(4), 2065–2073 (2014)
17. Lin, K.H., Lundquist, J.: Mate selection in cyberspace: the intersection of race, gender, and education. Am. J. Sociol. **119**(1), 183–215 (2013)

18. Newman, M.E.: Modularity and community structure in networks. Proc. Natl. Acad. Sci. U.S.A. **103**(23), 8577–8582 (2006). https://doi.org/10.1073/pnas.0601602103. http://www.ncbi.nlm.nih.gov/pubmed/16723398
19. Newman, M.E.J.: Finding community structure in networks using the eigenvectors of matrices.pdf. Phys. Rev. E **74**(3), 36,104 (2006)
20. Pereira, A.L.V., Hruschka, E.R.: Simultaneous co-clustering and learning to address the cold start problem in recommender systems. Knowl.-Based Syst. **82**, 11–19 (2015)
21. Pizzato, L., Chung, T., Rej, T., Koprinska, I., Yacef, K., Kay, J.: Learning user preferences in online dating. Technical report. University of Sydney, School of Information Technologies (2010). https://books.google.com/books?id=65DRZwEACAAJ
22. Pizzato, L., Rej, T., Chung, T., Koprinska, I., Kay, J.: RECON: a reciprocal recommender for online dating. In: Proceedings of the Fourth ACM Conference on Recommender Systems, TBA, pp. 207–214. ACM (2010). https://doi.org/10.1145/1864708.1864747. http://portal.acm.org/citation.cfm?id=1864708.1864747
23. Pizzato, L.A., Rej, T., Yacef, K., Koprinska, I., Kay, J.: Finding someone you will like and who won't reject you. In: User Modeling, Adaption and Personalization, pp. 269–280. Springer, Berlin (2011)
24. Skopek, J., Schulz, F., Blossfeld, H.P.: Who contacts whom? Educational homophily in online mate selection. Eur. Sociol. Rev. (2010). https://doi.org/10.1093/esr/jcp068
25. Smith, A., Anderson, M.: 5 Facts About Online Dating | Pew Research Center. http://www.pewresearch.org/fact-tank/2016/02/29/5-facts-about-online-dating/
26. Tu, K., Ribeiro, B., Jensen, D., Towsley, D., Liu, B., Jiang, H., Wang, X.: Online dating recommendations: matching markets and learning preferences. In: Proceedings of the Companion Publication of the 23rd International Conference on World Wide Web Companion, pp. 787–792. International World Wide Web Conferences Steering Committee (2014)
27. Xia, P., Ribeiro, B., Chen, C., Liu, B., Towsley, D.: A study of user behavior on an online dating site. In: Proceedings of the 2013 IEEE/ACM International Conference on Advances in Social Networks Analysis and Mining, pp. 243–247. ACM, New York (2013)
28. Xia, P., Jiang, H., Wang, X., Chen, C., Liu, B.: Predicting user replying behavior on a large online dating site. In: Proceedings of 8th International AAAI Conference on Weblogs and Social Media (2014)
29. Xia, P., Liu, B., Sun, Y., Chen, C.: Reciprocal recommendation system for online dating. In: Proceedings of the 2015 IEEE/ACM International Conference on Advances in Social Networks Analysis and Mining 2015, pp. 234–241. ACM, New York (2015)
30. Yan, B., Gregory, S.: Detecting community structure in networks using edge prediction methods. J. Stat. Mech. Theory Exp. **2012**(09), P09,008 (2012)
31. Yanxiang, L., Deke, G., Fei, C., Honghui, C.: User-based clustering with top-n recommendation on cold-start problem. In: 2013 Third International Conference on Intelligent System Design and Engineering Applications (ISDEA), pp. 1585–1589. IEEE, Washington (2013)
32. Yu, M., Zhao, K., Yen, J., Kreager, D.: Recommendation in reciprocal and bipartite social networks–a case study of online dating. In: Social Computing, Behavioral-Cultural Modeling and Prediction, pp. 231–239. Springer, Berlin (2013)
33. Yu, M., Zhang, X., Kreager, D.: New to online dating? learning from experienced users for a successful match. In: 2016 IEEE/ACM International Conference on Advances in Social Networks Analysis and Mining (ASONAM), pp. 467–470. IEEE, Washington (2016)
34. Zhao, K., Wang, X., Yu, M., Gao, B.: User recommendations in reciprocal and bipartite social networks–an online dating case study. IEEE Intell. Syst. **29**(2), 27–35 (2014)

Combining Feature Extraction and Clustering for Better Face Recognition

Salim Afra and Reda Alhajj

Abstract In this paper, we study the performance of face clustering approaches using different feature extraction techniques. This study will highlight best practices for handling faces of terrorists and criminals in an approach which we are working on to trace and red flag potential cases. Given as input images containing faces of people, face clustering divides them into K groups/clusters with each group containing images expected to represent almost the same person. Face clustering is very important, especially in forensic investigations where millions of images are available in crime scenes to be investigated. We study the performance of face clustering by first choosing different feature extraction techniques to capture information from faces. Feature extraction techniques are employed to check which face representation works better in describing faces as input to clustering algorithms. We also used Rank Order clustering algorithm which is known for its good accuracy when clustering face images along with other traditional clustering techniques. We evaluated the performance of feature extraction techniques and clustering algorithms using four datasets (JAFFE, AT&T, LFW, and YaleB); each imposing different challenges for face clustering with varying image environment and for datasets of different sizes. These datasets challenge clustering algorithms and feature extraction techniques in run time and clustering accuracy. Experimental results show the effectiveness of Rank Order clustering in terms of accuracy for small datasets while its run time performance degrades for larger datasets. K-means performed poorly on the LFW dataset. OpenFace performed the best in describing face images, especially on large datasets compared to other feature extraction techniques. The latter method reported high accuracy margin that is big and acceptable feature extraction time.

Keywords Face clustering · Face recognition · Feature extraction · Image processing

S. Afra · R. Alhajj (✉)
Department of Computer Science, University of Calgary, Calgary, AB, Canada
e-mail: salim.afra@ucalgary.ca; alhajj@ucalgary.ca

© Springer International Publishing AG, part of Springer Nature 2018   223
M. Kaya et al. (eds.), *Social Network Based Big Data Analysis and Applications*,
Lecture Notes in Social Networks, https://doi.org/10.1007/978-3-319-78196-9_11

1 Introduction

Face recognition involves detecting and verifying persons' identity by processing digital images and frames extracted from videos. Face recognition systems are becoming more popular due to rapidly advancing technology which made it affordable to capture and store a large number of images at low cost. They have various applications and benefits, including homeland security where video surveillance systems detect and recognize criminals or intruders. Video surveillance systems that are able to recognize people from a captured video stream are becoming more important, especially with incidents related to crimes, for example, the Boston marathon attack [12]. In such incidents, thousands of images are collected by video surveillance cameras and then inspectors analyze faces residing inside frames.

Clustering of people plays an important role during the investigation of crimes. In crowded areas, a large number of persons may pass in a specific location where a video surveillance system will keep on capturing image frames which can be in the order of millions. The same person may appear hundreds of times in frames which are not necessarily all consecutive. Thus, clustering of people will be perfect to apply for the following two main reasons:

- Filtering Data: By excluding images where no person is detected in surveillance camera frames.These images should be discarded during the investigation. The remaining images will be trimmed to concentrate only on persons appearing in frames, mainly their faces.
- Organizing Data: Here images of the same person in different location scenes will be identified to belong to the same cluster. This way, an investigator interested in tracking a specific person will only concentrate on a specific cluster and may proceed to identify and investigate other related suspects.

In order to cluster people in video frames, we use face as the identifier because a face is the most distinctive key to person's identity [5]. Clustering faces is a challenging process and dependable not only on the clustering algorithm invoked, but also on the feature extraction technique used. Both have several challenges to cope with. Feature representation challenges are inherited from limitations of visual features due to several factors, including low resolution of face images, changes in illumination between images, capturing a person from different viewpoints, cluttered background, etc., while clustering challenges may be attributed to the fact that the expected number of people in an input frame is not known in advance. This may cause a problem for some clustering techniques, mainly those which require a number of clusters as input. Another issue is that the number of images of different people is unbalanced. For instance, some people may appear in a few frames while others may exist in many frames; this aspect is challenging for some clustering techniques as discussed in Sect. 3.3.

Face recognition has recently received considerable attention as evident by the number of face recognition algorithms described in the literature. However, clustering of face images has not received enough attention yet. As a result, existing literature lacks on efforts which investigate an appropriate match between feature extraction techniques and clustering algorithms. Motivated by this, the work described in this paper evaluates the performance of various feature extraction and clustering techniques using a number of datasets of face images. By doing so, we seek to have a better idea on which feature extraction technique works better with a given clustering algorithm with respect to time and clustering accuracy.

The rest of the paper is organized as follows. Section 2 discusses existing literature related to face clustering. Section 3 describes the methodology used in face clustering along with feature extraction and clustering algorithms to be used in performance evaluation. Section 4 presents the datasets used in the evaluation. Section 5 includes the experiments and results. Section 6 is conclusions.

2 Related Work

Clustering has been applied in pattern recognition and has been successfully used in many different fields. Face clustering analysis has not received much attention as the clustering of faces depends not only on the clustering algorithm used but also on the feature extraction technique invoked. Different feature extraction techniques described in the literature can be applied as a preprocessing step for face clustering, but there is no widely accepted feature representation technique.

Several studies (e.g., [9, 24]) have evaluated the performance of a feature representation technique in association with a single clustering algorithm. For instance, Ho et al. [9] used Spectral clustering to evaluate the performance of local gradients with pixel intensity as a feature vector for face representation. Zhao et al. [24] used Hierarchical clustering to cluster photos in a personal gallery. They used a combination of features to represent a face image based on information extracted from face, body, and context information. Zhu et al. presented a new clustering algorithm specifically for face clustering [26]; it is called Rank Order distance clustering. Clustering is achieved by measuring the dissimilarity between two faces based on their neighborhood information. This is one of the clustering algorithms evaluated in this study. It is described in detail in Sect. 3.3.

Other studied investigated the effect of using feature representation techniques in combination with clustering algorithms. For instance, Heisele et al. [8] classified faces using Support Vector Machine (SVM) to evaluate three different feature representation techniques, namely, component-based method and two global methods for face recognition. In their evaluation, they used ROC curves of these feature representation techniques for formal and rotated faces. They determined that the component-based method outperformed the two global methods. While in [19] they present analysis similar to our study by checking the performance of different feature extraction techniques and clustering algorithms. They used component-based features and compared with a commercial face matcher. After extracting features from a face, they then apply three different clustering algorithms, namely, K-means, Spectral clustering, and Rank Order distance, which we have used in this study. They used two datasets for their experiments, namely, Pinellas County Sheriff's Office (PSCO) and Labeled Faces in the Wild (LFW) dataset. Their results show that the commercial face matcher outperformed the component-based for Rank Order clustering but they could not run the commercial face matcher on K-means or Spectral clustering because the feature vectors are not provided by the commercial product. They also show that Rank Order clustering performs better than K-means and Spectral clustering.

3   Methodology

As shown in Fig. 1, the general methodology of clustering faces consists of four main stages. The first step is to acquire a face dataset from any appropriate source which may be a video surveillance camera. The datasets we used for this purpose are mentioned in Sect. 4. After attaining the image collection, the next step is to preprocess and filter the images to concentrate only on faces of people. Then, feature extraction techniques are applied on the processed faces to get a feature vector of each image/face. The last step is to cluster the extracted feature vectors. More detail on each step is explained below.

3.1   Preprocessing

Face preprocessing of input images involves three major steps as depicted in Fig. 2.

1. **Detect Face Region:** The first step is to apply face detection over the image in order to get the face region of a person. By applying face detection, we eliminate extra details in the image and focus only on the face of the person. We have used dlib's implementation [11] of the Histograms of Oriented Gradients (HOG) face detection method as described in [6]. The output of this method is a face image of size 96×96 pixels.
2. **Face Landmark Detection:** After the face region is extracted, we compute face landmarks as shown in Fig. 2. We have used dlib's implementation of face pose estimation as presented in [10]. In their work, they have made an ensemble of

PRE-PROCESSING   FEATURE EXTRACTION   CLUSTERING

Fig. 1 Face clustering overall methodology

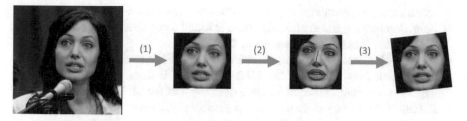

Fig. 2 Preprocessing steps example

regression trees to estimate landmark positions of a face from an image. They achieved high quality and fast predictions. The output from this method is 128 points that represent head pose.

3. **Face Alignment:** The last step performed is to align the head position straight with no rotation while keeping same eye, nose, and mouth position for all images. This step is important so that all faces are properly aligned because a slight variation in face alignment would be enough to trigger a false positive match with another person in the dataset. A face is aligned by using landmarks detected to put the eyes, mouth, and nose at a similar location for every image so that the features extracted for every face will have almost the same face position. This is done by doing affine transformation of faces with the help of landmarks to normalize and align faces at the same position.

3.2 Feature Extraction

After completing the preprocessing stage, face images become ready to extract their features. Extracting features of an image corresponds to building a feature vector which represents its important pixel information. These feature vectors are to be used in the clustering process. There are several feature extraction techniques that can be applied to the face. Currently, the top feature extraction technique is the one based on convolutional neural networks developed by Google's FaceNet [22]. In their work, they use up to 200 million private images of people to train a deep neural network to learn a feature vector of a face image and map it to a compact Euclidean space. Using this method, the similarity measure between two faces is simply the squared L2 distance between the two images.

In our analysis, we chose the following image feature extraction techniques. Then, we study the effect of using each of these feature extraction techniques with the clustering models.

- **SIFT** [15]: Scale-invariant feature transform (SIFT) method was developed by D. Lowe in 2004. SIFT extracts key-points of an image and then computes its descriptors. The algorithm to detect key-points involves four major steps: Scale-space extreme detection, key-point localization, orientation assignment, and key-point descriptor generation. Scale-space is found using an approximate Laplacian of Gaussian (LoG) with difference of Gaussian.

- **SURF** [4]: Speeded-Up Robust Features (SURF) method came out in 2006 as a speeded-up version of the SIFT algorithm. It does its speedup by using approximation algorithms to improve every step of the SIFT algorithm.
- **BRISK** [14]: Binary Robust Invariant Scalable Key points (BRISK) was developed in 2011 to make feature extraction effective and faster than previous methods such as SIFT and SURF. BRISK samples pattern out of concentric rings and then apply Gaussian smoothing. Building the descriptor is done by performing intensity comparisons.
- **DAISY** [23]: This method was developed in 2010. It depends on histograms of gradients like SIFT for key-point descriptor, but also uses a Gaussian weighting and circularly symmetrical kernel.
- **KAZE** [2]: Developed in 2012, this method analyzes and describes an image by operating in a nonlinear-scale space. The nonlinear-scale space is built efficiently by means of Additive Operator Splitting (AOS) schemes, which are stable for any step size and could be parallelized.
- **LBPH** [1]: The Local Binary Patterns Histograms (LBPH) feature extraction method can be described in the following steps:

  - Extract local features from images: This is done by not considering the whole image as a high-dimensional vector, instead describe only local features of a face. Features extracted this way will have low dimensions.
  - Summarize the local structure in an image by comparing each pixel with its neighborhood.
  - Take a pixel as center and threshold its neighbors accordingly.
  - Divide the LBP image into m local regions and extract a histogram from each. The corresponding feature vector of the fa
    ce is obtained by concatenating local histograms. These histograms are called Local Binary Patterns Histograms and the feature vectors of all images have the same size (size of the histogram).

- **OpenFace** [3]: The last feature extraction technique is OpenFace's implementation of FaceNet from Google. FaceNet yields the highest accuracy reported so far; the model and the data used in training remain private. For this purpose, OpenFace target was to implement the same neural network model of FaceNet and train it with 500k images from public datasets.

All these feature extraction methods, except for LBPH and OpenFace, identify local features in an image and calculate its descriptor as its feature vector. This local feature property of images would lead to feature vectors of different sizes. However, to classify faces, all images should have feature vectors of the same size. To overcome this problem, we follow the feature extraction process proposed in [20] and highlighted in Fig. 3. This feature extraction process has three main components.

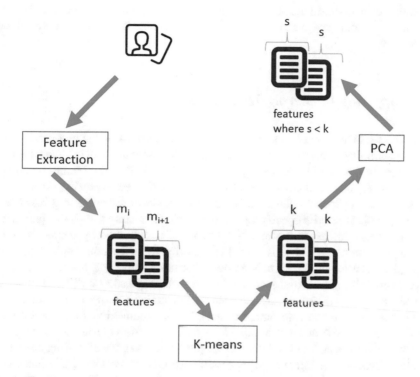

Fig. 3 Feature extraction process

1. **Image Feature Extraction:** The first step of the feature extraction process is to get as input face images and apply one of the general feature extraction methods described above to get corresponding feature vectors. These feature vectors may vary in size across different images.
2. K-**means Clustering:** After having the feature vectors of all face images, the next step is to map them into corresponding feature vectors of equal length. This is achieved by clustering the features using K-means to obtain K bins (where K is set to 200). As a result, every feature from the original feature vectors will be assigned a label of its cluster, in the range 1–200. These labels are used to build for each image a feature vector of fixed size K. This is done by iterating over original features of every image and increasing the ith entry of its new feature vector where i is the label of a given original feature.
3. **Principle Component Analysis:** The last step in the general feature extraction process is to reduce the dimensionality of the feature space produced based on K-means as described in step 2. Here, principal component analysis (PCA) is applied on the new feature vectors and leads to s features per image. PCA is a statistical method where given a set of feature vectors of possibly correlated variables, it converts correlated variables into a set of linearly uncorrelated

variables called principal components. This will eliminate unnecessary features from the feature space and will lead to more efficient clustering of the actual face images.

3.3 Applying Clustering Techniques

The last step in the process is to apply clustering algorithms on the produced feature vectors such that images of each person end up in a separate cluster. We have used in this study three of the clustering algorithms described in the literature, namely, K-**means** [7], **Spectral clustering** [18], and **Rank-Order clustering** [26].

K-means and Spectral clustering are the most widely used clustering algorithms. Both algorithms require specifying the number of clusters as an input parameter. This requires knowing in advance the number of people who appear in the video surveillance system, which is a serious restriction in several applications, for example, forensic investigation. Moreover, K-means suffer because the final result highly depends on the initial seeds of the clusters. This makes it difficult to handle clusters with varying density, size, and shape. Spectral clustering, on the other hand, can handle nonuniform distribution of data, but its complexity is high and usually performs poorly with noisy data [26]. Noise may come from the detection of faces in the various frames and will badly affect the performance of the clustering algorithm.

Rank-Order clustering method successfully tackles the problems associated with K-means and Spectral clustering. This method checks neighborhood of a face to determine its cluster. The method defines a new distance measure based on the dissimilarity in the neighborhood structure. Zhu et al. also claimed that their algorithm can handle nonuniform data distribution and it is robust to noise [26]. The algorithm has three major steps:

1. **Initialize Clusters:** Each face image forms a cluster on its own.
2. **Candidate Merging:** Compute the Rank Order distance between every pair of clusters C_i and C_j. If it is less than a threshold t, then mark the two clusters as a candidate merging pair.
3. **Transitive Merge:** Transitively merge all candidate merging pairs, then update the distance between clusters and loop back to the second step until no further merging is possible.

4 Datasets

We have used four datasets to evaluate the feature extraction techniques and the clustering algorithms described in the previous section. Each dataset has a different set of images to cluster, and each has its own challenges exhibited for face recognition.

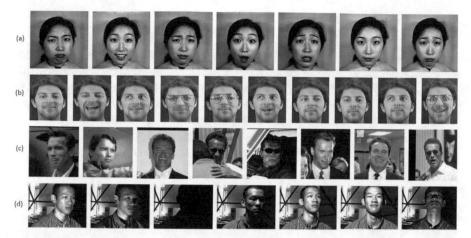

Fig. 4 (**a**) presents the JAFFE dataset, as seen all the images are taken in a controlled environment with the difference being the expression shown by the female. (**b**) presents the AT&T dataset where the face images are also taken in a controlled lab environment with difference in facial features and expressions for each person. (**c**) presents the LFW dataset; this dataset has a collection for each person images from the Web. As seen the images are not related to a scene and the image for a person is taken in different time frames (old/young). It presents a unique challenge to face recognition, as also other people can interfere in the image as shown in the sample. (**d**) presents the YaleB dataset, the dataset is taken in a controlled environment but the challenge here is with the illumination of each different image such that some images are unrecognizable even for humans

4.1 JAFFE Dataset

The Japanese Female Facial Expression (JAFFE) [16] database contains 213 images posted by 10 Japanese females. This dataset challenges face clustering by showing different facial expressions for each female in the dataset, these include happy, sad, angry, disgust, fear, surprise, and neutral. Examples from the JAFFE dataset are shown in Fig. 4a.

4.2 AT&T Face Dataset

The AT&T dataset [21] contains a set of faces collected at the University of Cambridge. The dataset contains 400 face images of 40 distinct persons, 10 images per person. The challenge for this dataset is to recognize people where each image was captured at different times with varying lightning scenes, different facial expressions, different facial details (glasses/no glasses), and different face alignments. An example of the dataset is shown in Fig. 4b.

4.3   LFW Dataset

The Labeled Faces in the Wild (LFW) dataset [25] contains 13,233 images of faces collected from the Web, representing 5749 persons. As shown in Fig. 4c, faces in this dataset are very challenging for face recognition algorithms as they were captured in the wild with varying conditions; the size of the clusters is varying in size and density because 1680 of the pictured persons have two or more distinct photos in the dataset, some have tens of images, and others have only one image. Given these challenges, this dataset could be classified as the hardest used in this study.

4.4   Extended YaleB Dataset

The Extended Yale Face Database B (Yale B) [13] is the largest dataset used in this study with 16,128 gray-scale images of 28 individuals. Every person has nine poses, where each pose has 65 images with a different facial expression or configuration. A sample of the dataset is shown in Fig. 4d.

5   Experiments and Results

In this section, we evaluate the performance of different clustering algorithms using several feature extraction methods on the four datasets mentioned in the previous section. We first present results of the preprocessing step, then we show performance of the feature extraction techniques and how this affects the clustering algorithms based on running time. Finally, we show the performance of the clustering algorithms for every feature extraction technique based on clustering accuracy. All experiments were run on a single machine with Intel Core i5-2400 CPU @ 3.1 GHz with 8 GB of RAM.

5.1   Face Preprocessing Results

Recall that the second step of the methodology described in Sect. 3.1 is to detect the face region in an image using an HOG descriptor for face detection. Table 1 reports for each dataset the original number of images the dataset has against the number of faces detected from our HOG face detector. All images from JAFFE dataset were successfully detected. We were able to apply preprocessing steps on them. This is expected with this dataset because it was captured in a controlled environment where the difference between the images is just facial expressions. However, face detection accuracy decreased drastically for AT&T and

Table 1 Preprocessing face detection accuracy

	Original	Detected	Percentage
JAFFE	213	213	100
AT&T	400	300	75
YaleB	16,380	90,371	55.17
LFW	13,233	13,176	99.54

YaleB datasets, scoring 75% and 55%, respectively. Even though both datasets were captured in a controlled environment, the HOG detector failed to identify faces with low illumination where their features can be hardly seen. This is why almost just 55% of the YaleB dataset has been detected; almost half the images of this dataset have low illumination. As for the LFW dataset which includes faces captured in an uncontrolled environment as shown in Fig. 4, we were able to detect almost the whole dataset with 99.54% detection accuracy. Having excellent accuracy in detecting the LFW dataset is very important because these images were taken in the wild and many different applications such as video surveillance systems capture images in a similar environment.

5.2 Feature Extraction Runtime

In this section, we report the running time results of the feature extraction process. It is very important to study the time required to extract the features and cluster the images. This is true because there are time-critical applications where time is an essential factor in deciding on the method to use and on acceptable clustering results.

As mentioned in Sect. 3.2, first each of **SIFT, SURF, KAZE, DAISY**, and **BRISK** extracts features of a face, then K-means and PCA are applied on the result to assign equal number of features for all images. To apply **LBPH** and **OpenFace**, we just have to get features from the given image.

Table 2 reports the run time for extracting features by the different extraction techniques for the listed datasets. Table 2 includes the following columns. "Extract Feature" time is common to all techniques and refers to the total time needed to extract features. "K-means" and "PCA" reveal the time needed by the feature extraction techniques. "Total" is the total time required by the features extraction techniques, K-means and PCA. As shown in Table 2, as the dataset size increases, the run time for the extraction techniques increases. From these results, it can be seen that SURF and KAZE are the fastest methods to extract features from images, closely followed by LBPH, and then OpenFace, while other methods take significantly more time to extract features for datasets. The difference margin is quite clear in case of **LFW** dataset where LBPH, SURF, DAISY, and OpenFace completed the process between 6 and 118 min, while KAZE needed almost 20 min, SIFT needed almost an hour and a half, and BRISK completed in around 2 and half hours. As detailed in the table, BRISK took so much time because of the extraction technique it uses, where the method needed considerable time to process a single

Table 2 Feature extraction
run time

	Extract features	K-means	PCA	Total
(a) JAFFE				
SIFT	0:00:13	0:00:23	0:00:01	0:00:37
SURF	0:00:02	0:00:01	0:00:01	0:00:05
KAZE	0:00:19	0:00:04	0:00:01	0:00:24
DAISY	0:00:03	0:00:02	0:00:01	0·00:06
BRISK	0:02:22	0:00:03	0:00:01	0:02:26
LBPH	0:00:05	–	–	0:00:05
OpenFace	0:00:09	–	–	0:00:09
(b) AT&T				
SIFT	0:00:19	0:00:55	0:00:02	0:01:15
SURF	0:00:04	0:00:02	0:00:01	0:00:07
KAZE	0:00:26	0:00:04	0:00:02	0:00:32
DAISY	0:00:05	0:00:03	0:00:01	0:00:09
BRISK	0:03:19	0:00:02	0:00:01	0:03:22
LBPH	0:00:08	–	–	0:00:08
OpenFace	0:00:19	–	–	0:00:19
(c) YaleB				
SIFT	0:07:51	0:43:37	0:00:15	0:51:43
SURF	0:02:16	0:00:49	0:00:15	0:03:20
KAZE	0:12:43	0:01:44	0:00:14	0:14:41
DAISY	0:02:22	0:01:44	0:00:12	0:04:18
BRISK	1:41:21	0:08:02	0:00:16	1:49:39
LBPH	0:03:30	–	–	0:03:30
OpenFace	0:07:19	–	–	0:07:19
(d) LFW				
SIFT	0:12:37	1:13:25	0:00:22	1:26:14
SURF	0:06:10	0:01:14	0:00:20	0:07:44
KAZE	0:17:30	0:02:27	0:00:20	0:20:17
DAISY	0:03:30	0:02:38	0:00:17	0:06:25
BRISK	2:29:25	0:06:41	0:00:23	2:36:29
LBPH	0:07:04	–	–	0:07:04
OpenFace	0:11:52	–	–	0:11:52

The table reports related to a dataset, a comparison of the run
time of the three different clustering techniques used in this
study. The results are shown in the format of H:MM:ss, where
H is hour, M is minutes, and s is seconds. Values marked as "–"
indicate that the clustering algorithm did not finish in a matter of
running for 1 day

image and get its features. While SIFT is not so slow in the extraction process, it
slows down when it moves to the K-means step to produce clusters. The reason
behind this is that SIFT generates a huge feature vector describing one image. So,
applying K-means on all features of every image requires a lot of time.

5.3 *Clustering Results*

In this section, we first show a study similar to that discussed in the previous section. Here, we analyze clustering time for different feature extraction techniques and clustering tools. As mentioned earlier, the importance of a clustering algorithm should compensate between run time and clustering accuracy.

Clustering run time is reported in Table 3. Run time is not reported for some clustering techniques, especially on large datasets. This is because we show only results for clustering algorithms that finished in at most 1 day time.

For the first three datasets described in Sect. 5.1, namely JAFFE and AT&T, corresponding results for all the feature extraction techniques are shown in Table 3a–c; these three datasets are the smallest in size in terms of the number of face images. Clustering run time for these three datasets is very fast; it is almost identical for K-means and Spectral clustering, taking almost a second each for all feature extraction techniques. On the other hand, the results show that Rank Order clustering is significantly slower than the other two clustering methods.

As reported in Table 3a for the JAFFE dataset, the performance of Rank Order clustering ranges between 12 and 18 s for all feature extraction techniques except for the LBPH method. However, it took 49 s to complete the clustering for the LBPH extraction technique. This is because the number of features generated by LBPH is larger than that of the others. Actually, LBPH constructs a feature histogram for every region in an image. The difference between the performance of LBPH and other methods can be clearly seen in Table 3b. To finish the clustering process, the other methods needed almost a minute, while LBPH took 2 min.

Concerning the two datasets, LFW and YaleB, K-means finished successfully on all the feature extraction techniques. Spectral clustering terminated successfully on the YaleB dataset but failed on LFW. On the other hand, Rank Order clustering couldn't finish running for both YaleB and LFW datasets. LBPH was again the slowest taking more than 3 h to complete on the LFW dataset, while running time of the other feature extraction techniques like K-means ranged from 1 to 2 min.

Concerning clustering accuracy, our aim is to determine the best feature extraction technique used for face recognition and the best performing clustering technique based on the extracted features. We evaluated the accuracy of the clustering results based on the confusion matrix of the set of class labels predicted by the clustering algorithm for which true values are known from the dataset. To calculate the adjacency matrix, we use external validity indices which were designed to measure the similarity between two partitions (predicted labels vs true labels). This method's confusion matrix as described in [17] represents the count of pairs of points based on whether they belong to the same cluster or not by considering the two partitions. For each pair in the predicted partition, we check whether these pairs have the same label or not and based on that populate the four entries in the confusion matrix, that is, true positive, true negative, false positive, and false negative counts.

Table 3 Clustering run time

	SIFT	SURF	KAZE	DAISY	BRISK	LBPH	OpenFace
(a) JAFFE							
K-means	0:00:01	0:00:01	0:00:01	0:00:01	0:00:01	0:00:01	0:00:01
Spectral	0:00:01	0:00:01	0:00:01	0:00:01	0:00:01	0:00:01	0:00:01
Rank order	0:00:17	0:00:15	0:00:12	0:00:18	0:00:17	0:00:49	0:00:12
(b) AT&T							
K-means	0:00:01	0:00:01	0:00:01	0:00:01	0:00:01	0:00:02	0:00:01
Spectral	0:00:02	0:00:02	0:00:02	0:00:02	0:00:02	0:00:03	0:00:03
Rank order	0:01:03	0:01:24	0:01:06	0:01:00	0:01:14	0:02:10	0:00:57
(c) YaleB							
K-means	0:00:03	0:00:02	0:00:02	0:00:02	0:00:02	0:01:36	0:00:02
Spectral	3:13:32	3:04:31	3:03:43	4:44:03	3:03:20	5:09:30	4:11:15
Rank order	–	–	–	–	–	–	–
(d) LFW							
K-means	0:01:40	0:01:03	0:01:36	0:00:23	0:01:13	3:07:41	0:02:22
Spectral	–	–	–	–	–	–	–
Rank order	–	–	–	–	–	–	–

The table reports results of a dataset, comparing run time of the three clustering techniques used in this study . The results are shown in the format of H:MM:ss where H is hour, M is minutes, and s is seconds. Values marked as "–" indicate that the clustering algorithm did not finish in a matter of running for 1 day

After getting the confusion matrix, we calculate precision and recall by considering images in each cluster produced by the clustering algorithm and compare them with the corresponding ground truth. A given cluster C may contain some face images which are indeed members of C based on the ground truth and some other face images which should have not been included in C. Precision is the proportion of face images that were correctly classified as members of C, that is, the number of correct faces in C divided by all faces in C. On the other hand, recall considers face images in the ground truth to determine their proportion correctly classified in C, that is, it is the number of face images correctly classified in C divided by the number of all face images which should have been classified in C according to the ground truth. We also calculate F-measure which is a summary statistic that combines precision and recall as given in Eq. (1).

$$F = 2 \times \frac{\text{Precision} \times \text{Recall}}{\text{Precision} + \text{Recall}} \tag{1}$$

Clustering accuracy results are presented in Table 4 for all the datasets and clustering algorithms applied on the feature extraction techniques. Figure 5 shows how clustering accuracy of Rank Order clustering varies for different threshold values.

Table 4 Clustering accuracy

	SIFT	SURF	KAZE	DAISY	BRISK	LBPH	OpenFace
(a) JAFFE							
K-means	**78.87%**	74.64%	68.3%	58.02%	55.46%	65.4%	78.46%
Spectral	73.88%	70.07%	75.6%	55.5%	50.9%	**80.16%**	63.28%
Rank order	64.68% (22)	66.4% (26)	74.14% (22)	63.59% (26)	40.64% (21)	71.61% (25)	**82.14%** **(21)**
(b) AT&T							
K-means	56.05%	35.16%	40.91%	48.1%	23.71%	37.55%	**83.83%**
Spectral	50.33%	34.62%	35.82%	39.65%	18.59%	35.53%	**77.45%**
Rank order	52.15% (24)	41.03% (19)	44.82% (16)	49.76% (13)	26.22% (15)	57.22% (11)	**94.78%** **(19)**
(c) YaleB							
K-means	6.63%	4.73%	5.5%	5.49%	4.72%	8.46%	**65.84%**
Spectral	6.66%	4.83%	5.53%	6.64%	4.29%	6.65%	**75.11%**
Rank order	–	–	–	–	–	–	–
(d) LFW							
K-means	0.16%	0.01%	0.16%	0.17%	0.05%	0.24%	**1.41%**
Spectral	–	–	–	–	–	–	–
Rank order	–	–	–	–	–	–	–

The table reports results of a dataset, comparing run time of the three clustering techniques used in this study. The results are shown in the format of H:MM:ss where H is hour, M is minutes, and s is seconds. Values marked as "–" indicate that the clustering algorithm did not finish in a matter of running for 1 day

Table 4 reports accuracy results obtained by applying the different clustering algorithms on each feature extraction technique mentioned above. Values of best clustering accuracy for each clustering algorithm are shown in bold font. Table 4a shows the results obtained for JAFFE dataset. The best result in this table is obtained by using Rank Order clustering with the OpenFace feature extraction technique (82%). This is followed by using Spectral clustering on the LBPH method (80%). We can conclude from this table that the OpenFace technique has on average the highest accuracy across the three different clustering techniques. The second best-performing method is LBPH, while BRISK performed on average the lowest. This conclusion is further supported in Table 4b where the difference margin becomes clear as the dataset size increases.

The best clustering results belong to OpenFace using Rank Order clustering (94.78%). In this table and all following tables, it is possible to notice that the best results for *K*-means, Spectral, and Rank Order clustering have been obtained using the OpenFace technique. Another conclusion which could be noticed from this table is that Rank Order clustering gave on average the best results compared to the other clustering algorithms.

Table 4c shows clustering results for the YaleB dataset which has almost 9000 images after preprocessing. In the table, we miss the values of Rank Order

Clustering which gave the best results in the previous table. This is because Rank Order clustering could not finish in 1 day. The results show a big gap when using OpenFace compared to the other feature extraction techniques, reaching 60%.

Spectral clustering with OpenFace reached 75% while the best result shown by the other methods is LBPH (8%). Table 4d reports accuracy results for the LFW dataset where only K-means finished in a day. OpenFace reported best results; its accuracy is just 1.41%, while accuracy for the other techniques ranges from 0.01% to 0.24%. This is because the LFW dataset has imbalanced cluster sizes, and algorithms like K-means would fail when applied on this kind of dataset.

Figure 5 shows Rank Order clustering results in terms of precision, recall, and F-measure for JAFFE, ATT&T, and ATT&T Filtered image sets. This figure shows that having low threshold will result in high precision almost 100%, while recall is low. And while the threshold of the clustering increases, precision starts to decrease, recall gets higher, and F-measure always goes up to a certain threshold then back down with the margin of precision and recall getting high.

This can be explained as follows, for a low threshold, most images will be merged together to belong to the same cluster. Having high precision means most retrieved images have the same label. Low recall means a smaller percentage of images from a target cluster have been retrieved. As the threshold increases, few images will be considered to belong to the same cluster. The number of clusters will increase such that person B will have a cluster with some noise, that is, recall will increase, while some of person A images will end up in other clusters (due to false positives) leading to lower precision.

6 Conclusions

We have performed a study on different feature extraction techniques to determine which one is better to use in Face clustering. In addition to feature extraction techniques, we also used three clustering algorithms to see which clustering algorithm performs the best on features extracted by the feature extraction techniques.

For this purpose, we have used four datasets, each with its own challenges in the face recognition problem and with varying sizes. We did our experiments on a single machine and noted down the results based on both run time and clustering accuracy. From the experiments and results, we concluded that the best clustering algorithm is Rank Order clustering, but due to its time complexity we could not manage to run it for large datasets. As for time complexity, K-means run time is massively better than Rank Order clustering and Spectral clustering. Spectral clustering is even faster than Rank Order clustering. Concerning the best feature extraction technique, it was OpenFace that performed the best when used with Rank Order clustering. Not only accuracy levels were great by good margins, also feature extraction time was acceptable and in a good range compared to the other techniques.

As for future work, we would like to enhance Rank Order clustering by using approximation methods to construct a K nearest neighbor (K-NN) graph instead of

Fig. 5 Plots of all the rankorder results with varying thresholds

calculating the distance between a given image and all other images in the same dataset. Also we want to use more enhanced hardware to perform parallel execution of this procedure, thus making it faster. This way, it would become possible to faster produce clustering results for LFW and YaleB datasets.

References

1. Ahonen, T., Hadid, A., Pietikainen, M.: Face description with local binary patterns: application to face recognition. IEEE Trans. Pattern Anal. Mach. Intell. **28**(12), 2037–2041 (2006)
2. Alcantarilla, P.F., Bartoli, A., Davison, A.J.: Kaze features. In: Computer Vision–ECCV 2012, pp. 214–227. Springer, Berlin (2012)
3. Amos, B., Ludwiczuk, B., Satyanarayanan, M.: Openface: a general-purpose face recognition library with mobile applications. Technical report, CMU-CS-16-118, CMU School of Computer Science (2016)
4. Bay, H., Tuytelaars, T., Gool, L.V.: Surf: speeded up robust features. In: Computer Vision–ECCV 2006, pp. 404–417. Springer, Berlin (2006)
5. Bruce, V., Young, A.: Understanding face recognition. Br. J. Psychol. **77**(3), 305–327 (1986)
6. Dalal, N., Triggs, B.: Histograms of oriented gradients for human detection. In: 2005 IEEE Computer Society Conference on Computer Vision and Pattern Recognition (CVPR'05), vol. 1, pp. 886–893. IEEE, Washington (2005)
7. Hartigan, J.A., Wong, M.A.: Algorithm as 136: a k-means clustering algorithm. J. R. Stat. Soc. Ser. C (Appl. Stat.) **28**(1), 100–108 (1979)
8. Heisele, B., Ho, P., Poggio, T.: Face recognition with support vector machines: global versus component-based approach. In: Proceedings of the Eighth IEEE International Conference on Computer Vision, 2001. ICCV 2001, vol. 2, pp. 688–694. IEEE, Washington (2001)
9. Ho, J., Yang, M.-H., Lim, J., Lee, K.-C., Kriegman, D.: Clustering appearances of objects under varying illumination conditions. In: Proceedings of the 2003 IEEE Computer Society Conference on Computer Vision and Pattern Recognition, 2003, vol. 1, pp. I–11. IEEE, Washington (2003)
10. Kazemi, V., Sullivan, J.: One millisecond face alignment with an ensemble of regression trees. In: Proceedings of the IEEE Conference on Computer Vision and Pattern Recognition, pp. 1867–1874 (2014)
11. King, D.E.: Dlib-ml: a machine learning toolkit. J. Mach. Learn. Res. **10**(July), 1755–1758 (2009)
12. Klontz, J.C., Jain, A.K.: A case study of automated face recognition: the boston marathon bombings suspects. Computer **46**(11), 91–94 (2013)
13. Lee, K.-C., Ho, J., Kriegman, D.J.: Acquiring linear subspaces for face recognition under variable lighting. IEEE Trans. Pattern Anal. Mach. Intell. **27**(5), 684–698 (2005)
14. Leutenegger, S., Chli, M., Siegwart, R.Y.: Brisk: binary robust invariant scalable keypoints. In: 2011 IEEE International Conference on Computer Vision, pp. 2548–2555. IEEE, Washington (2011)
15. Lowe, D.G.: Distinctive image features from scale-invariant keypoints. Int. J. Comput. Vis. **60**(2), 91–110 (2004)
16. Lyons, M.J., Akamatsu, S., Kamachi, M., Gyoba, J., Budynek, J.: The Japanese female facial expression (jaffe) database (1998)
17. Milligan, G.W., Cooper, M.C.: A study of the comparability of external criteria for hierarchical cluster analysis. Multivar. Behav. Res. **21**(4), 441–458 (1986)
18. Ng, A.Y., Jordan, M.I., Weiss, Y., et al.: On spectral clustering: analysis and an algorithm. Adv. Neural Inf. Process. Syst. **2**, 849–856 (2002)
19. Otto, C., Klare, B., Jain, A.K.: An efficient approach for clustering face images. In: 2015 International Conference on Biometrics, pp. 243–250. IEEE, Washington (2015)
20. Puttemans, S., Howse, J., Hua, Q., Sinha, U.: Opencv 3 Blueprints: Expand Your Knowledge of Computer Vision by Building Amazing Projects with Opencv 3. Packt Publishing, Birmingham (2015)
21. Samaria, F.S., Harter, A.C.: Parameterisation of a stochastic model for human face identification. In: Proceedings of the Second IEEE Workshop on Applications of Computer Vision, 1994, pp. 138–142. IEEE, Washington (1994)

22. Schroff, F., Kalenichenko, D., Philbin, J.: Facenet: a unified embedding for face recognition and clustering. In: Proceedings of the IEEE Conference on Computer Vision and Pattern Recognition, pp. 815–823 (2015)
23. Tola, E., Lepetit, V., Fua, P.: Daisy: an efficient dense descriptor applied to wide-baseline stereo. IEEE Trans. Pattern Anal. Mach. Intell. **32**(5), 815–830 (2010)
24. Zhao, M., Teo, Y.W., Liu, S., Chua, T.-S., Jain, R.: Automatic person annotation of family photo album. In: International Conference on Image and Video Retrieval, pp. 163–172. Springer, Berlin (2006)
25. Zhu, X., Ramanan, D.: Face detection, pose estimation, and landmark localization in the wild. In: 2012 IEEE Conference on Computer Vision and Pattern Recognition, pp. 2879–2886. IEEE, Washington (2012)
26. Zhu, C., Wen, F., Sun, J.: A rank-order distance based clustering algorithm for face tagging. In: 2011 IEEE Conference on Computer Vision and Pattern Recognition, pp. 481–488. IEEE, Washington (2011)

Index

Printed in the United States
By Bookmasters